Entomology: Study of Insects

Entomology: Study of Insects

Edited by
Jael Payne

⊟ Larsen & Keller
www.larsen-keller.com

Entomology: Study of Insects
Edited by Jael Payne
ISBN: 978-1-63549-108-1 (Hardback)

Larsen & Keller

Published by Larsen and Keller Education,
5 Penn Plaza,
19th Floor,
New York, NY 10001, USA

Cataloging-in-Publication Data

Entomology : study of insects / edited by Jael Payne.
    p. cm.
Includes bibliographical references and index.
ISBN 978-1-63549-108-1
1. Entomology. 2. Insects. I. Payne, Jael.
QL463 .E58 2017
595.7--dc23

This book contains information obtained from authentic and highly regarded sources. All chapters are published with permission under the Creative Commons Attribution Share Alike License or equivalent. A wide variety of references are listed. Permissions and sources are indicated; for detailed attributions, please refer to the permissions page. Reasonable efforts have been made to publish reliable data and information, but the authors, editors and publisher cannot assume any responsibility for the vailidity of all materials or the consequences of their use.

Trademark Notice: All trademarks used herein are the property of their respective owners. The use of any trademark in this text does not vest in the author or publisher any trademark ownership rights in such trademarks, nor does the use of such trademarks imply any affiliation with or endorsement of this book by such owners.

The publisher's policy is to use permanent paper from mills that operate a sustainable forestry policy. Furthermore, the publisher ensures that the text paper and cover boards used have met acceptable environmental accreditation standards.

Printed and bound in the United States of America.

For more information regarding Larsen and Keller Education and its products, please visit the publisher's website www.larsen-keller.com

Table of Contents

Preface   **VII**

Chapter 1   **Introduction to Entomology**   **1**

Chapter 2   **Branches of Entomology**   **8**
- i. Coleopterology   8
- ii. Myrmecology   9
- iii. Medical Entomology   14
- iv. Forensic Entomology   17
- v. Economic Entomology   29

Chapter 3   **Insect and its Classification**   **34**
- i. Insect   34
- ii. Beetle   63
- iii. Hemiptera   90
- iv. Termite   104
- v. Lepidoptera   128
- vi. Bee   159
- vii. Ant   177
- viii. Grasshopper   200
- ix. Caddisfly   214
- x. Wasp   217

Chapter 4   **Insect Ecology: An Overview**   **233**
- i. Insect Ecology   233
- ii. Insect Migration   233
- iii. Insect Hotel   237

Chapter 5   **Various Equipments used in Entomology**   **241**
- i. Insect Trap   241
- ii. Bottle Trap for Insects   244
- iii. Flight Interception Trap   247
- iv. Killing Jar   248
- v. Malaise Trap   249
- vi. Moth Trap   250
- vii. Pheromone Trap   252
- viii. Pitfall Trap   253
- ix. Tullgren Funnel   254
- x. Electrical Penetration Graph   255

Chapter 6 **Applications of Entomology**   **258**
            i.   Biological Pest Control   258
            ii. Beekeeping   270
            iii. Sericulture   294
            iv. Biomimetics   297

           Permissions

           Index

Preface

Entomology is that branch of zoology which deals with the study of insects. It studies the structure, classification, biological systems, nature, habitats, etc. of insects in detail. This book attempts to understand the multiple topics that fall under the discipline of entomology and how such concepts have practical applications and effects in the ecosystem. It picks up individual branches and explains their need and contribution in the context of the growth of this area. The textbook also explains the various practices that highlight the conservation of the environment. Coherent flow of topics, student-friendly language and extensive use of examples make this text an invaluable source of knowledge.

A short introduction to every chapter is written below to provide an overview of the content of the book:

Chapter 1 - The branch of zoology that studies insects is known as entomology. Insects include earthworms, snails, slugs and myriapods etc. Entomology overlaps with a number of topics such as molecular genetics, biomechanics, biochemistry and ecology. The chapter on entomology offers an insightful focus, keeping in mind the complex subject matter; **Chapter 2** - Entomology has a number of branches; some of these are coleopterology, myrmecology, medical entomology, forensic entomology and economic entomology. Coleopterology is the study of beetles whereas the study of ants is known as myrmecology. This text is a compilation of the various branches of entomology that form an integral part of the broader subject matter; **Chapter 3** - Insects are the most diverse form of animals and represent more than a million species. Insects can be found in every habitat, although only a small number is found in the oceans. This chapter lists insects such as, beetle, hemiptera, termite, lepidopter, bee, ant, grasshopper and wasp. The chapter serves as a source to understand the major categories related to insects; **Chapter 4** - Insect ecology is concerned with the interaction of insects with the surrounding environment. Insects play a number of roles in the environment, such as pest control, pollination and soil turning and aeration. They are crucial for the biodiversity present on Earth. This section is an overview of the subject matter incorporating all the major aspects of insect ecology; **Chapter 5** - Insects are reduced for certain number of reasons; insect traps are used to directly decrease populations of insects. The mechanisms involved in trapping insects vary, as different insects are attracted to different objects. Some of the equipments used in trapping insects are bottle traps for insects, flight interception traps, malaise traps, moth traps and pitfall traps. The aspects elucidated in this text are of vital importance, and provide a better understanding of entomology; **Chapter 6** - Methods used for controlling pests are known as biological control.

Beekeeping, sericulture and biomimetics are some of the applications of entomology that have been elucidated in this text. The diverse applications of entomology in the current scenario have been thoroughly discussed in the following chapter;

I extend my sincere thanks to the publisher for considering me worthy of this task. Finally, I thank my family for being a source of support and help.

Editor

Introduction to Entomology

The branch of zoology that studies insects is known as entomology. Insects include earthworms, snails, slugs and myriapods etc. Entomology overlaps with a number of topics such as molecular genetics, biomechanics, biochemistry and ecology. The chapter on entomology offers an insightful focus, keeping in mind the complex subject matter. -

A phasmid, mimicking a leaf

Entomology is the scientific study of insects, a branch of zoology. In the past the term "insect" was more vague, and historically the definition of entomology included the study of terrestrial animals in other arthropod groups or other phyla, such as arachnids, myriapods, earthworms, land snails, and slugs. This wider meaning may still be encountered in informal use.

Like several of the other fields that are categorized within zoology, entomology is a taxon-based category; any form of scientific study in which there is a focus on insect-related inquiries is, by definition, entomology. Entomology therefore overlaps with a cross-section of topics as diverse as molecular genetics, behavior, biomechanics, biochemistry, systematics, physiology, developmental biology, ecology, morphology, paleontology, mathematics, anthropology, robotics, agriculture, nutrition, forensic science, and more.

At some 1.3 million described species, insects account for more than two-thirds of all known organisms, date back some 400 million years, and have many kinds of interactions with humans and other forms of life on earth.

History

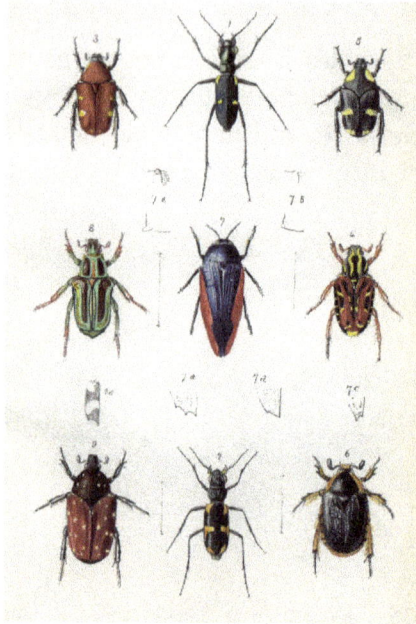

Plate from Transactions of the Entomological Society, 1848

Entomology is rooted in nearly all human cultures from prehistoric times, primarily in the context of agriculture (especially biological control and beekeeping), but scientific study began only as recently as the 16th century.

William Kirby is widely considered as the father of Entomology. In collaboration with William Spence, he published a definitive entomological encyclopedia, *Introduction to Entomology*, regarded as the subject's foundational text. He also helped to found the Royal Entomological Society in London in 1833, one of the earliest such societies in the world; earlier antecedents, such as the Aurelian society date back to the 1740s.

Entomology developed rapidly in the 19th and 20th centuries, and was studied by large numbers of people, including such notable figures as Charles Darwin, Jean-Henri Fabre, Vladimir Nabokov, Karl von Frisch (winner of the 1973 Nobel Prize in Physiology or Medicine), and two-time Pulitzer Prize winner E. O. Wilson.

Identification of Insects

Most insects can easily be recognized to order such as Hymenoptera (bees, wasps, and ants) or Coleoptera (beetles). However, insects other than Lepidoptera (butterflies and moths) are typically identifiable to genus or species only through the use of Identification keys and Monographs. Because the class Insecta contains a very large number of species (over 330,000 species of beetles alone) and the characteristics separating them are unfamiliar, and often subtle (or invisible without a microscope), this is often very difficult even for a specialist. This has led to the development of automated species identification systems targeted on insects, for example, Daisy, ABIS, SPIDA and Drawwing

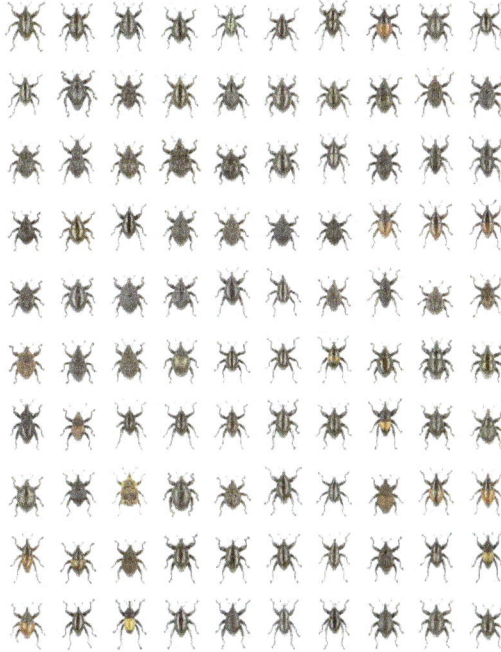

These 100 Trigonopterus species were described simultaneously using DNA barcoding

Insect identification is an increasingly common hobby, with butterflies and dragonflies being the most popular.

In Pest Control

In 1994 the Entomological Society of America launched a new professional certification program for the pest control industry called The Associate Certified Entomologist (ACE). To qualify as a "true entomologist" an individual would normally require an advanced degree, with most entomologists pursuing their PhD. While not true entomologists in the traditional sense, individuals who attain the ACE certification may be referred to as ACEs, Amateur entomologists, Associate entomologists or –more commonly– Associate-Certified Entomologists.

Taxonomic Specialization

Many entomologists specialize in a single order or even a family of insects, and a number of these subspecialties are given their own informal names, typically (but not always) derived from the scientific name of the group:

Part of a large beetle collection

- Coleopterology - beetles

- Dipterology - flies

- Hemipterology - true bugs

- Isopterology - termites

- Lepidopterology - moths and butterflies

- Melittology (or *Apiology*) - bees

- Myrmecology - ants

- Orthopterology - grasshoppers, crickets, etc.

- Trichopterology - caddis flies

- Vespology - Social wasps

Organizations

Like other scientific specialties, entomologists have a number of local, national, and international organizations. There are also many organizations specializing in specific subareas.

- Amateur Entomologists' Society

- Deutsches Entomologisches Institut

- Entomological Society of America

- Entomological Society of Canada

- Entomological Society of Japan

- International Union for the Study of Social Insects

- Netherlands Entomological Society

- Royal Belgian Entomological Society

- Royal Entomological Society of London

- Société entomologique de France

Museums

Here is a list of selected museums which contain very large insect collections.

Africa

- Natal Museum, Pietermaritzburg, South Africa

Europe

- Muséum national d'histoire naturelle, Paris, France

- Museum für Naturkunde, Berlin, Germany

- Natural History Museum, Budapest Hungarian Natural History Museum

- Natural History Museum, Geneva

- Natural History Museum, Leiden, the Netherlands

- Natural History Museum, London, United Kingdom

- Natural History Museum, Oslo Norway

- Natural History Museum, St. Petersburg Zoological Collection of the Russian Academy of Science

- Naturhistorisches Museum, Vienna, Austria

- Oxford University Museum of Natural History, Oxford

- Royal Museum for Central Africa, Brussels, Belgium

- Swedish Museum of Natural History, Stockholm, Sweden

- The Bavarian State Collection of Zoology Zoologische Staatssammlung München

- World Museum Liverpool, the *Bug House*

United States

- Academy of Natural Sciences of Philadelphia

- American Museum of Natural History, New York City

- Auburn University Museum of Natural History, Auburn, Alabama
- Audubon Insectarium, New Orleans
- Bohart Museum of Entomology, Davis, California
- California Academy of Sciences, San Francisco
- Carnegie Museum of Natural History, Pittsburgh
- Essig Museum, Berkeley, California
- Field Museum of Natural History, Chicago
- Florida Museum of Natural History, University of Florida, Gainesville, Florida
- Illinois Natural History Survey, Champaign, Illinois
- J. Gordon Edwards Museum, San Jose, California
- Museum of Comparative Zoology, Cambridge, Massachusetts
- Natural History Museum of Los Angeles County, Los Angeles
- National Museum of Natural History, Washington, D.C.
- New Mexico State University Arthropod Museum
- North Carolina State University Insect Museum, Raleigh, North Carolina
- Peabody Museum of Natural History, New Haven, Connecticut
- The National Museum of Play, Rochester, N.Y.
- Texas A&M University, College Station, Texas
- University of Minnesota, St. Paul campus (UMSP), Minnesota
- University of Kansas Natural History Museum, Lawrence, Kansas
- University of Nebraska State Museum, Lincoln, Nebraska
- University of Missouri Enns Entomology Museum, University of Missouri, Columbia, Missouri

Canada

- Canadian Museum of Nature, Ottawa
- Canadian National Collection of Insects, Arachnids and Nematodes, Ottawa, Ontario

- E.H. Strickland Entomological Museum, University of Alberta, Edmonton, Alberta

- Lyman Entomological Museum, Macdonald Campus of McGill University, Sainte-Anne-de-Bellevue, Quebec

- Montreal Insectarium, Montreal, Quebec

- Newfoundland Insectarium, Reidville, Newfoundland and Labrador

- Royal Alberta Museum, Edmonton, Alberta

- Royal Ontario Museum, Toronto

- University of Guelph Insect Collection, Guelph, Ontario

- Victoria Bug Zoo, Victoria, British Columbia

References

- Liddell, Henry George and Robert Scott (1980). A Greek-English Lexicon (Abridged Edition). United Kingdom: Oxford University Press. ISBN 0-19-910207-4.

- Chapman, A. D. (2006). Numbers of living species in Australia and the World. Canberra: Australian Biological Resources Study. pp. 60pp. ISBN 978-0-642-56850-2.

- Antonio Saltini, Storia delle scienze agrarie, 4 vols, Bologna 1984-89, ISBN 88-206-2412-5, ISBN 88-206-2413-3, ISBN 88-206-2414-1, ISBN 88-206-2415-X

- NMSU Entomology Plant Pathology; Weed science. "New Mexico State University Arthropod Museum". Retrieved 2013-07-15.

2

Branches of Entomology

Entomology has a number of branches; some of these are coleopterology, myrmecology, medical entomology, forensic entomology and economic entomology. Coleopterology is the study of beetles whereas the study of ants is known as myrmecology. This text is a compilation of the various branches of entomology that form an integral part of the broader subject matter.

Coleopterology

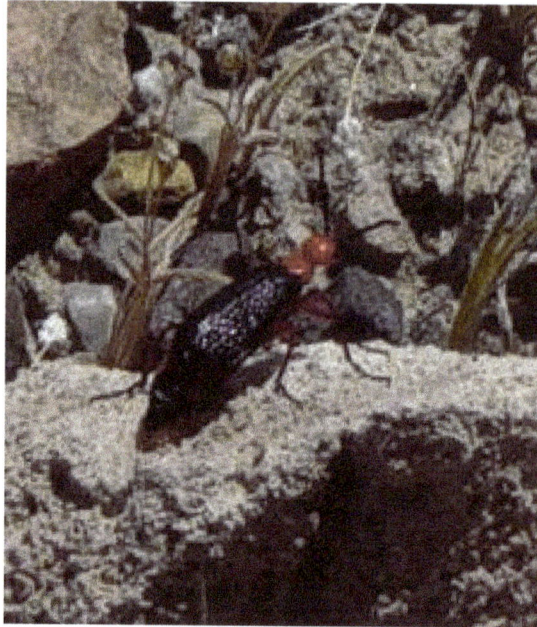

Leatherhead Beetle above Mesquite Springs in Death Valley

Coleopterology is a branch of entomology, the scientific study of beetles of the order Coleoptera). Practitioners are termed coleopterists. Coleopterists have formed organizations to facilitate the study of beetles. Among these is The Coleopterists Society, an international organization based in the United States. Such organizations may have both professionals and amateurs as members, interested in beetles. When speaking informally, coleopterists sometimes refer to their study as "beetling".

Literature

- *J. Cooter & M. V. L. Barclay, ed. (2006). A Coleopterist's Handbook. Amateur Entomological Society. ISBN 0-900054-70-0.*

- *E. Reitter, ed. (1908–1917). Fauna Germanica. The beetles of the German Reich..*

- *A. Horion, ed. (1941–1974). faunistics the Central European beetles.*

- *H. Joy; KW Harde; GA Lohse, eds. (1964–1983). The beetles of Central Europe Goecke & Evers, Krefeld. ISBN 3-334-61035-7..*

- KW Harde, F. Severa: The Cosmos Beetle leader Franckh, Stuttgart, 1981. ISBN 3-440-04881-0 .

- Wolfgang Willner: Pocket Dictionary of beetles of Central Europe Source & Meier, Wiebelsheim 2013. ISBN 978-3-494-01451-7 .

Myrmecology

Meat eater ant feeding on honey

Myrmecology is a branch of entomology focusing on the scientific study of ants. Some early myrmecologists considered ant society as the ideal form of society and sought to find solutions to human problems by studying them. Ants continue to be a model of choice for the study of questions on the evolution of social systems because of their complex and varied forms of eusociality. Their diversity and prominence in ecosystems also has made them important components in the study of biodiversity and conservation. Recently, ant colonies are also studied and modeled for their relevance in machine learning, complex interactive networks, stochasticity of encounter and interaction networks, parallel computing, and other computing fields.

History

The word myrmecology was coined by William Morton Wheeler (1865–1937), although human interest in the life of ants goes back further, with numerous ancient folk references. The earliest scientific thinking based on observation of ant life was that of Auguste Forel (1848–1931), a Swiss psychologist who initially was interested in ideas of instinct, learning, and society. In 1874 he wrote a book on the ants of Switzerland, *Les fourmis de la Suisse*, and he named his home *La Fourmilière* (the ant colony). Forel's early studies included attempts to mix species of ants in a colony. He noted polydomy and monodomy in ants and compared them with the structure of nations.

Wheeler looked at ants in a new light, in terms of their social organization, and in 1910 he delivered a lecture at Woods Hole on the "The Ant-Colony as an Organism," which pioneered the idea of superorganisms. Wheeler considered trophallaxis or the sharing of food within the colony as the core of ant society. This was studied using a dye in the food and observing how it spread in the colony.

Some, such as Horace Donisthorpe, worked on the systematics of ants. This tradition continued in many parts of the world until advances in other aspects of biology were made. The advent of genetics, ideas in ethology and its evolution led to new thought. This line of enquiry was pioneered by E. O. Wilson, who founded the field termed as sociobiology.

Interdisciplinary Application

Ants often are studied by engineers for biomimicry and by network engineers for more efficient networking. It is not known clearly how ants manage to avoid congestions and how they optimize their movements to move in most efficient ways without a central authority that would send out orders. There already have been many applications in structure design and networking that have been developed from studying ants, but the efficiency of human-created systems is still not close to the efficiency of ant colonies.

List of Notable Myrmecologists

- Ernest André (1838–1911), French entomologist

- Thomas Borgmeier (1892–1975), German-Brazilian theologian and entomologist
- William L. Brown, Jr. (1922–1997), American entomologist
- Giovanni Cobelli (1849–1937), Italian entomologist, director of the Rovereto museum
- Arthur Charles Cole, Jr. (1908–1955), American entomologist
- Walter Cecil Crawley, British entomologist
- William Steel Creighton (1902–1973)
- Horace Donisthorpe (1870–1951), British myrmecologist, named several new species
- Carlo Emery (1848–1925), Italian entomologist
- Johan Christian Fabricius (1745–1808), Danish entomologist, student of Linnaeus
- Auguste-Henri Forel (1848–1931), Swiss myrmecologist, studied brain structure of humans and ants
- Émil August Goeldi (1859–1917), Swiss-Brazilian naturalist and zoologist
- William Gould (1715–1799), described by Horace Donisthorpe as "the father of British myrmecology"
- Robert Edmond Gregg (1912–1991), American entomologist
- Thomas Caverhill Jerdon (1811–1872), British physician, zoologist and botanist
- Walter Wolfgang Kempf (1920–1976), Brazilian myrmecologist
- Heinrich Kutter (1896–1990), Swiss myrmecologist
- Nicolas Kusnezov also as Nikolaj Nikolajevich Kuznetsov-Ugamsky (1898–1963)
- Pierre André Latreille (1762–1833) French entomologist
- Sir John Lubbock (the 1st Lord and Baron Avebury) (1834–1913), wrote on hymenoptera sense organs
- William T. Mann (1886–1960), American entomologist
- Gustav Mayr (1830–1908), Austrian entomologist and professor in Pest and Vienna, specialised in Hymenoptera
- Carlo Menozzi also as Carlo Minozzi (1892–1943), Italian entomologist

- Wilhelm Nylander (1822–1899), Finnish botanist, briologist, micologist, entomologist and myrmecologist

- Basil Derek Wragge-Morley (1920–1969), research included genetics, social behaviour of animals, and the behaviour of agricultural pests

- Fergus O'Rourke (1923– 2010), Irish zoologist

- Julius Roger (1819–1865), German physician, entomologist and folklorist

- Felix Santschi (1872–1940), Swiss entomologist

- Theodore Christian Schneirla (1902–1968), American animal psychologist

- Frederick Smith (1805–1879), worked in the zoology department of the British Museum from 1849, specialising in the Hymenoptera

- Roy R. Snelling (1934–2008), American entomologist credited with many important finds of rare or new ant species

- Erich Wasmann (1859–1931), Austrian entomologist

- Neal Albert Weber (1908–2001), American myrmecologist

- John Obadiah Westwood (1805–1893), English entomologist and archaeologist also noted for his artistic talents

- William Morton Wheeler (1865–1937), curator of invertebrate zoology in the American Museum of Natural History, described many new species

Contemporary Myrmecologists

- Donat Agosti, Swiss entomologist

- Cesare Baroni Urbani, Swiss ant taxonomist

- Murray S. Blum (1929–), American chemical ecologist, an expert on pheromones

- Barry Bolton, English ant taxonomist

- Alfred Buschinger, German myrmecologist

- Henri Cagniant, French myrmecologist

- John S. Clark, Scottish myrmecologist

- Cedric Alex Collingwood, British entomologist

- Mark Amidon Deyrup, American myrmecologist

- Francesc Xavier Espadaler i Gelabert, Spanish (Catalan) myrmecologist, specialist in Mediterranean and Macaronesian ants and in invasive species

- Deborah Gordon (1955–), studies ant colony behavior and ecology

- William H. Gotwald, Jr., American entomologist

- Michael J. Greene studies interactions between chemical cues and behavior patterns

- Bert Hölldobler (1936–), Pulitzer Prize winning German myrmecologist

- Laurent Keller (1961–), Swiss evolutionary biologist and myrmecologist

- John E. Lattke

- John T. Longino, American entomologist

- Mark W. Moffett (1958–), American entomologist and photographer

- Corrie S. Moreau, American evolutionary biologist and entomologist, wrote on evolution and diversification of ants

- Justin Orvel Schmidt, American entomologist, studies the chemical and behavioral defenses of ants, wasps, and arachnids

- Bernhard Seifert, German entomologist

- Steven O. Shattuck, American-Australian entomologist

- Marion R. Smith, American entomologist

- Robert W. Taylor, Australian myrmecologist

- Alberto Tinaut Ranera, Spanish myrmecologist

- Walter R. Tschinkel, American myrmecologist

- James C. Trager, American myrmecologist

- Gary J. Umphrey, American biostatistician and myrmecologist

- Philip S. Ward, American entomologist

- Edward Osborne Wilson (1929–), Pulitzer Prize winning American myrmecologist, revolutionized the field of sociobiology

Related Terms

- Myrmecochorous (adj.) dispersed by ants

- Myrmecophagous (adj.) feeding on ants

- Myrmecophile (n.) an organism that habitually shares an ant nest, *myrmecoph-*

ilous (adj.), *myrmecophily* (n.)

- Myrmidons (n.) ant-men in Metamorphoses and in Homer's Iliad, where they are Achilles' warriors

Medical Entomology

Aedes albopictus

The discipline of medical entomology, or public health entomology, and also *veterinary entomology* is focused upon insects and arthropods that impact human health. Veterinary entomology is included in this category, because many animal diseases can "jump species" and become a human health threat, for example, bovine encephalitis. Medical entomology also includes scientific research on the behavior, ecology, and epidemiology of arthropod disease vectors, and involves a tremendous outreach to the public, including local and state officials and other stake holders in the interest of public safety, finally in current situation related to one health approach mostly health policy makers recommends to widely applicability of medical entomology for disease control efficient and best fit on achieving development goal and to tackle the newly budding zoonotic diseases. Thoughtful to have and acquaint with best practice of Med. Entomologist to tackle the animal and public health issues together with controlling arthropods born diseases by having Medical Entomologists' the right hand for bringing the healthy world [Yon w].

Medical Entomologists are employed by private and public universities, private industries, and federal, state, and local government agencies, including all three branches of the US military - who hire medical entomologists to protect the troops from infectious diseases that can be transmitted by arthropods. Historically, during wars, more people have died due to insect-transmitted diseases, than to all the battle injuries combined.

Medical entomologists are also hired by chemical companies - to help develop new pesticides which will effectively decrease insect pest populations while simultaneously protecting the health of the public.

Public health entomology has seen a huge surge in interest since 2005, due to the resurgence of the bed bug, *Cimex lectularius*.

Insects of Medical Importance

Medical entomologists work in the public health arena, dealing with insects (and other arthropods) that parasitize people, bite, sting, and/or vector disease.

Personal Pests

Some personal pests of may vector pathogens:Lice, Fleas, Bedbugs, Ticks, Scabies mites

The Housefly

The housefly is a very common and cosmopolitan species which transmits diseases to man. The organisms of both amoebic and bacillary dysenteries are picked up by flies from the faeces of infected people and transferred to clean food either on the fly's hairs or by the fly vomiting during feeding. Typhoid germs may be deposited on food with the fly's faeces. The house fly cause the spread of yaws germs by carrying them from a yaws ulcer to an ordinary sore. Houseflies also transmit poliomyelitis by carrying the virus from infected faeces to food or drink. Cholera and hepatitis are sometimes fly-borne. Other diseases carried by houseflies are Salmonella, tuberculosis, anthrax, and some forms of ophthalmia. They carry over 100 pathogens and transmit some parasitic worms. The flies in poorer and lower-hygiene areas usually carry more pathogens. Some strains have become immune to most common insecticides.

The Cockroach

Cockroaches carry disease-causing organisms (typically gastroenteritis) as they forage. Cockroach excrement and cast skins also contain a number of allergens causing responses such as, watery eyes,skin rashes, congestion of nasal passages and asthma.

Biting Insects

Pathogen infection transmitted by insect or other arthropod vectors.

Diseases carried by insects and other arthropod vectors affect more than 700 million people every year, and are considered the most sensitive to climatic and environment conditions.(WHO)

Major Insect-born Disease

- Dengue fever - Vectors: *Aedes aegypti* (main vector) *Aedes albopictus* (minor vector) threatens -50 million people are infected by dengue annually, 25,000 die. Threatens 2.5 billion people in more than 100 countries.

- Malaria - Vectors: *Anopheles* mosquitoes - 500 million become severely ill with malaria every year and more than 1 million die.

- Leishmaniasis - Vectors: species in the genus *Lutzomyia* in the New World and *Phlebotomus* in the Old World. Two million people infected.

- Bubonic plague - Principal vector: *Xenopsylla cheopis* At least 100 flea species can transmit plague. Re-emerging major threat several thousand human cases per year.High pathogenicity and rapid spread.

- Sleeping sickness - Vector: Tsetse fly, not all species. Sleeping sickness threatens millions of people in 36 countries of sub-Saharan Africa (WHO)

- Typhus - Vectors: mites, fleas and body lice 16 million cases a year, resulting in 600,000 deaths annually.

- Wuchereria bancrofti - most common vectors: the mosquito species: *Culex, Anopheles, Mansonia*, and *Aedes*; affects over 120 million people.

- Yellow Fever - Principal vectors: *Aedes simpsoni, A. africanus*, and *A. aegypti* in Africa, species in *Haemagogus* genus in South America, and species in *Sabethes* genus in France -200,000 estimated cases of yellow fever (with 30,000 deaths) per year.

Minor

- Ross River fever - Vector: Mosquitoes, main vectors *A. vigilax, Aedes camptorhynchus*, and *Culex annulirostris*

- Barmah Forest Virus - Vector: Known vectors *Culex annulirostris, Ocleratus vigilax* and *O. camptorhynchus* and *Culicoides marksi*

- Kunjin encephalitis (mosquitoes)

- Murray Valley encephalitis virus (MVEV) - Major mosquito vector: *Culex annulirostris.*

- Japanese encephalitis - Several mosquito vectors, the most important being *Culex tritaeniorhynchus.*

- West Nile virus - Vectors: vary according to geographical area; in the United States *Culex pipiens* (Eastern US), *Culex tarsalis* (Midwest and West), and *Cu-*

lex quinquefasciatus (Southeast) are the main vectors.

- Lyme disease - Vectors: several species of the genus *Ixodes*

- Alkhurma virus (KFDV) - Vector: tick

- Kyasanur forest disease - Vector: *Haemaphysalis spinigera*

- Brugia timori filariasis - Primary vector: *Anopheles barbirostris*

- Babesia - Vector *Ixodes* ticks.

- Carrion's disease - Vectors: sandflies of the genus *Lutzomyia*.

- Chagas disease - Vector: assassin bugs of the subfamily *Triatominae*. The major vectors are species in the genera *Triatoma, Rhodnius*, and *Panstrongylus*.

- Chikungunya - Vectors: *Aedes* mosquitoes

- Human ewingii ehrlichiosis - Vector: *Amblyomma americanum*

- Human granulocytic ehrlichiosis - Vector: *Ixodes scapularis*

- Rift Valley Fever (RVF) - Vectors: fleas in the genera *Aedes* and *Culex*

- Scrub typhus - Vector: Chigger

- Loa loa filariasis - Vector: *Chrysops* sp.

Forensic Entomology

Forensic entomology is a science that is based on the scientific study of the invasion and succession pattern of arthropods with their developmental stages of different species found on the decomposed cadavers during legal investigations. Forensic entomology is the application and study of insect and other arthropod biology to criminal matters. It also involves the application of the study of arthropods, including insects, arachnids, centipedes, millipedes, and crustaceans to criminal or legal cases. It is primarily associated with death investigations; however, it may also be used to detect drugs and poisons, determine the location of an incident, and find the presence and time of the infliction of wounds. Forensic entomology can be divided into three subfields: urban, stored-product and medico-legal/medico-criminal entomology.

History

Historically, there have been several accounts of applications for, and experimentation with, forensic entomology. The concept of forensic entomology dates back to at least the

13th century. However, only in the last 30 years has forensic entomology been systematically explored as a feasible source for evidence in criminal investigations. Through their own experiments and interest in arthropods and death, Sung Tzu, Francesco Redi, Bergeret d'Arbois, Jean Pierre Mégnin and the physiologist Hermann Reinhard have helped to lay the foundations for today's modern forensic entomology.

Sung Tzu

Sung Tzu (also known as Sung Tz'u) was a Judicial Intendant who lived in China 1188-1251 AD. In 1247 AD Sung Tzu wrote a book entitled *Washing Away of Wrongs* as a handbook for coroners. In this book Sung Tzu depicts several cases in which he took notes on how a person died and elaborates on probable causes. He explains in detail on how to examine a corpse both before and after burial. He also explains the process of how to determine a probable cause of death. The main purpose of this book was to be used as a guide for other investigators so they could assess the scene of the crime effectively. His level of detail in explaining what he observed in all his cases laid down the fundamentals for modern forensic entomologists and is the first recorded account in history of someone using forensic entomology for judicial means.

Francesco Redi

In 1668, Italian physician Francesco Redi disproved the theory of spontaneous generation. The accepted theory of Redi's day claimed that maggots developed spontaneously from rotting meat. In an experiment, he used samples of rotting meat that were either fully exposed to the air, partially exposed to the air, or not exposed to air at all. Redi showed that both fully and partially exposed rotting meat developed fly maggots, whereas rotting meat that was not exposed to air did not develop maggots. This discovery completely changed the way people viewed the decomposition of organisms and prompted further investigations into insect life cycles and into entomology in general.

Bergeret d'Arbois

Dr. Louis François Etienne Bergeret (1814–1893) was a French hospital physician, and was the first to apply forensic entomology to a case. In a case report published in 1855 he stated a general life cycle for insects and made many assumptions about their mating habits. Nevertheless, these assumptions led him to the first application of forensic entomology in an estimation of post-mortem interval (PMI). His report used forensic entomology as tool to prove his hypothesis on how and when the person had died.

Hermann Reinhard

The first systematic study in forensic entomology was conducted in 1881 by Hermann Reinhard, a German medical doctor who played a vital role in the history of forensic entomology. He exhumed many bodies and demonstrated that the development of many different types of insect species could be tied to buried bodies. Reinhard conducted his first study in east Germany, and collected many Phorid flies from this initial study. He also concluded that the development of only some of the insects living with corpses underground were associated with them, since there were 15-year-old beetles who had little direct contact with them. Reinhard's works and studies were used extensively in further forensic entomology studies.

Jean Pierre Mégnin

Jean Pierre Mégnin (1828–1905), an army veterinarian, published many articles and books on various subjects including the books *Faune des Tombeaux* and *La Faune des Cadavres*, which are considered to be among the most important forensic entomology books in history. In his second book he did revolutionary work on the theory of predictable waves, or successions of insects onto corpses. By counting numbers of live and dead mites that developed every 15 days and comparing this with his initial count on the infant, he was able to estimate how long that infant was dead.

In this book he asserted that exposed corpses were subject to eight successional waves, whereas buried corpses were only subject to two waves. Mégnin made many great discoveries that helped shed new light on many of the general characteristics of decaying flora and fauna. Mégnin's work and study of the larval and adult forms of insect families found in cadavers sparked the interest of future entomologists and encouraged more research in the link between arthropods and the deceased, and thereby helped to establish the scientific discipline of forensic entomology.

Forensic Entomology Subfields

Urban Forensic Entomology

Urban forensic entomology typically concerns pests infestations in buildings gardens or that may be the basis of litigation between private parties and service providers such as landlords or exterminators. Urban forensic entomology studies may also indicate the appropriateness of certain pesticide treatments and may also be used in stored products cases where it can help to determine chain of custody, when all points of possible infestation are examined in order to determine who is at fault.

Stored-product Forensic Entomology

Stored-product forensic entomology is often used in litigation over insect infestation or contamination of commercially distributed foods.

Medico-legal Forensic Entomology

Medicolegal forensic entomology covers evidence gathered through arthropod studies at the scenes of murder, suicide, rape, physical abuse and contraband trafficking. In murder investigations it deals with which insects eggs appear, their location on the body and in what order they appear. This can be helpful in determining a post mortem interval (PMI) and location of a death in question. Since many insects exhibit a degree of endemism (occurring only in certain places), or have a well-defined phenology (active only at a certain season, or time of day), their presence in association with other evidence can demonstrate potential links to times and locations where other events may have occurred. Another area covered by medicolegal forensic entomology is the relatively new field of entomotoxicology. This particular branch involves the utilization of entomological specimens found at a scene in order to test for different drugs that may have possibly played a role in the death of the victim.

Invertebrate Types

Scorpionflies

Scorpionflies (order Mecoptera) were the first insects to arrive at a donated human cadaver observed (by the entomologist Natalie Lindgren) at the Southeast Texas Applied Forensic Science Facility near Huntsville, Texas, and remained on the corpse for one and a half days, outnumbering flies during that period. The presence of scorpionflies thus indicates that a body must be fresh.

Flies

Flies (order Diptera) are often first on the scene. They prefer a moist corpse for their offspring (maggots) to feed on. The most significant types of fly include:

- Blow flies – Family Calliphoridae- Flies in this family are often metallic in appearance and between 10 and 14 mm in length. In addition to the name blowfly, some members of this family are known as blue bottle fly, cluster fly, greenbottles, or black blowfly. A characteristic of the blow-fly is its 3-segmented antennae. Hatching from an egg to the first larval stage takes from eight hours to one day. Larvae have three stages of development (called instars); each stage is separated by a molting event. Worldwide, there are 1100 known species of blowflies, with 228 species in the Neotropics, and a large number of species in Africa and Southern Europe. The most common area to find Calliphoridae species are in the countries of India, Japan, Central America, and in the southern United States. The typical habitat for blow-flies are temperate to tropical areas that provide a layer of loose, damp soil and litter where larvae may thrive and pupate. The forensic importance of this fly is that it is the first insect to come in contact with carrion because they have the ability to smell death from up to ten miles (16 km) away.

- Flesh flies – Family Sarcophagidae- Most flesh flies breed in carrion, dung, garbage, or decaying material, but a few species lay their eggs in the open wounds of mammals; hence their common name. Characteristics of the flesh-fly is its 3-segmented antennae. Most holarctic Sarcophagidae vary in size from 4 to 18 mm in length (Tropical species can be larger) with black and gray longitudinal stripes on the thorax and checkering on the abdomen. Flesh-flies, being viviparous, frequently give birth to live young on corpses of human and other animals, at any stage of decomposition, from newly dead through to bloated or decaying (though the latter is more common).

Flesh fly on decomposing flesh

- House fly – Family Muscidae- is the most common of all flies found in homes, and indeed one of the most widely distributed insects; it is often considered a pest that can carry serious diseases. The adults are 6–9 mm long. Their thorax is gray, with four longitudinal dark lines on the back. The underside of their abdomen is yellow, and their whole body is covered with hair. Each female fly can lay up to 500 eggs in several batches of about 75 to 150 eggs. Genus *Hydrotaea* are of particular forensic importance.

- Cheese flies – Family Piophilidae - Most are scavengers in animal products and fungi. The best-known member of the family is *Piophila casei*. It is a small fly, about four mm (1/6 inch) long, found worldwide. This fly's larva infests cured meats, smoked fish, cheeses, and decaying animals and is sometimes called the cheese skipper for its leaping ability. Forensic entomology uses the presence of Piophila casei larvae to help estimate the date of death for human remains. They do not take up residence in a corpse until three to six months after death. The adult fly's body is black, blue-black, or bronze, with some yellow on the head, antennae, and legs. The wings are faintly iridescent and lie flat upon the fly's abdomen when at rest. At four mm (1/6 inch) long, the fly is one-third to one-half as long as the common housefly.

- Coffin flies – Phoridae

- Lesser corpse flies – Sphaeroceridae

- Lesser house flies – Fanniidae

- Black scavenger flies – Sepsidae

- Sun flies - Heleomyzidae

- Black soldier fly - Stratiomyidae - have potential for use in forensic entomology. The larvae are common scavengers in compost heaps, are found in association with carrion, can be destructive pests in honey bee hives, and are used in manure management (for both house fly control and reduction in manure volume). The larvae range in size from 1/8 to 3/4 of an inch (3 to 19 millimeters). The adult fly is a mimic, very close in size, color, and appearance to the organ pipe mud dauber wasp and its relatives.

- Phoridae–Humpbacked flies
  Larvae feed on decaying bodies. Some species can burrow to a depth of 50 cm over 4 days. Important in buried bodies.

- Non-biting midges - Chironomidae - these flies have a complex life cycle. While adults are terrestrial and phytophagous, larvae are aquatic and detritivorous. Immature instars have been used as forensic markers in several cases where submerged corpses were found.

Beetles

Beetles (Order Coleoptera) are generally found on the corpse when it is more decomposed. In drier conditions, the beetles can be replaced by moth flies (Psychodidae).

- Rove beetles – family Staphylinidae – are elongate beetles with small elytra (wing covers) and large jaws. Like other beetles inhabiting carrion, they have fast larval development with only three larval stages. Creophilus species are common predators of carrion, and since they are large, are a very visible component of the fauna of corpses. Some adult Staphylinidae are early visitors to a corpse, feeding on larvae of all species of fly, including the later predatory fly larvae. They lay their eggs in the corpse, and the emerging larvae are also predators. Some species have a long development time in the egg, and are common only during the later stages of decomposition. Staphylinids can also tear open the pupal cases of flies, to sustain themselves at a corpse for long periods.

- Hister beetles – family Histeridae. Adult histerids are usually shiny beetles (black or metallic-green) which have an introverted head. The carrion-feeding species only become active at night when they enter the maggot-infested part of the corpse to capture and devour their maggot prey. During daylight they hide under the corpse unless it is sufficiently decayed to enable them to hide inside it. They have fast larval development with only two larval stages. Among the first beetles to arrive at a corpse are Histeridae of the genus *Saprinus*. *Saprinus* adults feed on both the larvae and pupae of blowflies, although some have a preference for fresh pupae. The adults lay their eggs in the corpse, inhabiting it in the later stages of decay.

- Carrion beetles – family Silphidae- Adult Silphidae have an average size of about 12 mm. They are also referred to as burying beetles because they dig and bury small carcasses underground. Both parents tend to their young and exhibit communial breeding. The male carrion beetle's job in care is to provide protection for the breed and carcass from competitors.

- Ham beetles – family Cleridae

- Carcass beetles – family Trogidae

- Skin/hide beetles – family Dermestidae. Hide beetles are important in the final stages of decomposition of a carcass. The adults and larvae feed on the dried skin, tendons and bone left by fly larvae. Hide beetles are the only beetle with the enzymes necessary for breaking down keratin, a protein component of hair.

- Scarab beetles – family Scarabaeidae- Scarab beetles may be any one of around 30,000 beetle species worldwide that are compact, heavy-bodied and oval in shape. The flattened plates, which each antenna terminates, are fitted together to form a club. The outer edges of the front legs may also be toothed or scalloped. Scarab beetles range from 0.2 to 4.8 in (5.1 to 121.9 mm) in length. These species are known for being one of the heaviest insect species.

- Sap beetles – family Nitidulidae

Mites

Many mites (class Acari, not insects) feed on corpses with *Macrocheles* mites common in the early stages of decomposition, while Tyroglyphidae and Oribatidae mites such as Rostrozetes feed on dry skin in the later stages of decomposition.

Nicrophorus beetles often carry on their bodies the mite *Poecilochirus* which feed on fly eggs. If they arrive at the corpse before any fly eggs hatch into maggots, the first eggs are eaten and maggot development is delayed. This may lead to incorrect PMI estimates. *Nicrophorus* beetles find the ammonia excretions of blowfly maggots toxic, and the *Poecilochirus* mites, by keeping the maggot population low, allow *Nicrophorus* to occupy the corpse.

Moths

Moths (order Lepidoptera) specifically clothes-moths – Family Tineidae – are closely related to butterflies. Most species of moth are nocturnal, but there are crepuscular and diurnal species. Moths feed on mammalian hair during their larval stages and may forage on any hair that remains on a body. They are amongst the final animals contributing to the decomposition of a corpse.

Wasps, Ants, and Bees

Wasps, ants, and bees (order Hymenoptera) are not necessarily necrophagous. While some feed on the body, some are also predatory, and eat the insects feeding on the body. Bees and wasps have been seen feeding on the body during the early stages. This may cause problems for murder cases in which larval flies are used to estimate the post mortem interval since eggs and larvae on the body may have been consumed prior to the arrival on scene of investigators.

Factors

Moisture Levels

Rain and humidity levels in the area where the body is found can affect the time for insect development. In most species, large amounts of rain will indirectly cause slower development due to drop in temperature. Light rain or a very humid environment, by acting as an insulator, will permit a greater core temperature within the maggot mass, resulting in faster development.

Submerged Corpses

M. Lee Goff, a noted and well respected forensic entomologist, was assigned to a case involving the discovery of a decomposing body found on a boat half a mile from shore. Upon collection of the maggot mass, only one insect, *Chrysomya megacephala*, was discovered. He concluded that the water barrier accounted for the scarcity of other flies. He also noted that flies will not attempt to trek across large bodies of water unless there is a substantially influential attractant.

In addition, the amount of time a maggot mass has been exposed to salt water can affect its development. From the cases Goff observed he found that if subjected for more than 30 minutes, there was a 24hour developmental delay. Unfortunately, not many more studies have been conducted and thus a specific amount of delay time is difficult to estimate.

Sun Exposure

"Because insects are cold-blooded animals, their rate of development is more or less dependent on ambient temperature." Bodies exposed to large amounts of sunlight will heat up, giving the insects a warmer area to develop, reducing their development time. An experiment conducted by Bernard Greenberg and John Charles Kunich with the use of rabbit carcasses to study accumulation of degree days found that with temperature ranging in the mid 70s to high 80s the amount of developmental time for maggots was significantly reduced.

In contrast, bodies found in shaded areas will be cooler, and insects will require longer growth periods. In addition, if temperatures reach extreme levels of cold, insects instinctively know to prolong their development time in order to hatch into a more accepting and viable climate in order to increase the chance of survival and reproduction.

Air Exposure

Hanged bodies can be expected to show their own quantity and variety of flies. Also, the amount of time flies will stay on a hanged body will vary in comparison to one found on the ground. A hanged body is more exposed to air and thus will dry out faster leaving less food source for the maggots.

As the body begins to decompose, a compilation of fluids will leak to the ground. In this area most of the expected fauna can be found. Also, it is more likely that rove beetles and other non-flying insects will be found here instead of directly on the body. Fly maggots, initially deposited on the body, may also be found below.

Geography

According to Jean Pierre Mégnin's book *La Faune des Cadavres* there are eight distinct faunal successions attracted to a corpse. While most beetles and flies of forensic importance can be found worldwide, a portion of them are limited to a specific range of habitats. It is forensically important to know the geographical distribution of these insects is order to determine information such as post mortem interval or whether a body has been moved from its original place of death.

Calliphoridae is arguably the most important family concerning forensic entomology given that they are the first to arrive on the corpse. The family's habitat ranges into the southern portion of the United States. However, while *Chrysomya rufifaces*, the hairy maggot blow fly, is part of the Calliphoridae family and is widespread, it is not prevalent in the Southern California, Arizona, New Mexico, Louisiana, Florida, or Illinois regions.

Flesh flies fall under the family Sacrophagidae and generally arrive to a corpse following Calliphoridae. However, as previously mentioned they are capable of flying in the rain. This key advantage enables them to occasionally reach a body before Calliphoridae overall effecting the maggot mass that will be discovered. Flesh flies are globally distributed including habitats in the United States, Europe, Asia, and the Middle East.

Beetles are representative of the order Coleoptera which accounts for the largest of the insect orders. Beetles are very adaptive and can be found in almost all environments with the exception of Antarctica and high mountainous regions. The most diverse beetle fauna can be found in the tropics. In addition, beetles are less submissive to temperatures. Thus, if a carcass has been found in cold temperatures, the beetle will be prevalent over Calliphoridae.

Weather

Various weather conditions in a given amount of time cause certain pests to invade human households. This is because the insects are in search of food, water, and shelter. Damp weather causes reproduction and growth enhancement in many insect types, especially when coupled with warm temperatures. Most pests concerned at this time are ants, spiders, crickets, cockroaches, ladybugs, yellowjackets, hornets, mice, and rats. When conditions are dry, the deprivation of moisture outside drives many pests inside searching for water. While the rainy weather increases the numbers of insects, this dry weather causes pest invasions to increase. The pests most commonly known during dry conditions are scorpions, ants, pillbugs, millipedes, crickets, and spiders. Extreme drought does kill many populations of insects, but also drives surviving insects to invade more often. Cold temperatures outside will cause invasions beginning in the late summer months and early fall. Box elder bugs, cluster flies, ladybugs, and silverfish are noticed some of the most common insects to seek the warm indoors.

Modern Techniques

Many new techniques have been developed and are used in order to more accurately gather evidence, or reevaluate at old information. The use of these newly developed techniques and evaluations have become relevant in litigation and appeals. Forensic entomology not only uses arthropod biology, but it pulls from other sciences, introducing fields like chemistry and genetics, exploiting their inherent synergy through the use of DNA in forensic entomology.

Scanning Electron Microscopy

Fly larvae and fly eggs are used to aid in the determination of a PMI. In order for the data to be useful the larvae and eggs must be identified down to a species level to get an accurate estimate for the PMI. There are many techniques currently being developed to differentiate between the various species of forensically important insects. A study in 2007 demonstrates a technique that can use scanning electron microscopy (SEM) to identify key morphological features of eggs and maggots. Some of the morphological differences that can help identify the different species are the presence/absence of anastomosis, the presence/absence of holes, and the shape and length of the median area.

The SEM method provides an array of morphological features for use in identifying fly eggs; however, this method does have some disadvantages. The main disadvantage is that it requires expensive equipment and can take time to identify the species from which the egg originated, so it may not be useful in a field study or to quickly identify a particular egg. The SEM method is effective provided there is ample time and the proper equipment and the particular fly eggs are plentiful. The ability to use these morphological differences gives forensic entomologists a powerful tool that can help with estimating a post mortem interval, along with other relevant information, such as whether the body has been disturbed post mortem.

Potassium Permanganate Staining

When scanning electron microscopy is not available, a faster, lower cost technique is potassium permanganate staining. The collected eggs are rinsed with a normal saline solution and placed in a glass petri dish. The eggs are soaked in a 1% potassium permanganate solution for one minute and then dehydrated and mounted onto a slide for observation. These slides can be used with any light microscope with a calibrated eyepiece to compare various morphological features. The most important and useful features for identifying eggs are the size, length, and width of the plastron, as well as the morphology of the plastron in the area around the micropyle. The various measurements and observations when compared to standards for forensically important species are used to determine the species of the egg.

Mitochondrial DNA

In 2001, a method was devised by Jeffrey Wells and Felix Sperling to use mitochondrial DNA to differentiate between different species of the subfamily Chrysomyinae. This is particularly useful when working to determine the identity of specimens that do not have distinctive morphological characteristics at certain life stages.

Mock Crime Scenes

A valuable tool that is becoming very common in the training of forensic entomologists is the use of mock crime scenes using pig carcasses. The pig carcass represents a human body and can be used to illustrate various environmental effects on both arthropod succession and the estimate of the post mortem interval.

Gene Expression Studies

Although physical characteristics and sizes at various instars have been used to estimate fly age, a more recent study has been conducted to determine the age of an egg based on the expression of particular genes. This is particularly useful in determining developmental stages that are not evidenced by change in size; such as the egg or pupa and where only a general time interval can be estimated based on the duration of the particular developmental stage. This is done by breaking the stages down into smaller units separated by predictable changed in gene expression. Three genes were measured in an experiment with *Drosophila melanogaster*: bicoid (bcd), slalom (sll), and chitin synthase (cs). These three genes were used because they are likely to be in varied levels during different times of the egg development process. These genes all share a linear relationship in regards to age of the egg; that is, the older the egg is the more of the particular gene is expressed.

However, all of the genes are expressed in varying amounts. Different genes on different loci would need to be selected for another fly species. The genes expressions are mapped in a control sample to formulate a developmental chart of the gene expression at certain time intervals. This chart can then be compared to the measured values of gene expression to accurately predict the age of an egg to within two hours with a high confidence level. Even though this technique can be used to estimate the age of an egg, the feasibility and legal acceptance of this must be considered for it to be a widely utilized forensic technique. One benefit of this would be that it is like other DNA-based techniques so most labs would be equipped to conduct similar experiments without requiring new capital investment. This style of age determination is in the process of being used to more accurately find the age of the instars and pupa; however, it is much more complicated, as there are more genes being expressed during these stages. The hope is that with this and other similar techniques a more accurate PMI can be obtained.

Insect Activity Case Study

A preliminary investigation of insect colonization and succession on remains in New Zealand revealed the following results on decay and insect colonization.

Open Field Habitat

This environment had a daily average maximum temperature of 19.4 °C (66.9 °F) and a daily minimum temperature of 11.1 °C (52.0 °F). The average rainfall for the first 3 weeks in this environment was 3.0 mm/day. Around days 17–45, the body began to start active decay. During this stage, the insect successions started with *Calliphora stygia*, which lasted until day 27. The larvae of *Chrysomya rufifacies* were present between the day 13 and day 47. The *H. rostrata*, larvae of *Lucilia sericata*, Psychodidae family, and sylvicola were found to occur relatively late in the body's decay.

Coastal Sand-dune Habitat

This environment had an average daily maximum temperature of 21.4 °C (70.5 °F) and minimum of 13.5 °C (56.3 °F). The daily average rainfall was recorded as 1.4 mm/day for the first 3 weeks. The post-decay time interval, beginning at day six after death and ending around day 15 after death, is greatly reduced from the average post-decay time, due to the high average temperature of this environment. Insects obtained late in the post-active stage include the *Calliphora quadrimaculata*, adult Sphaeroceridae, Psychodidae and Piophilidae (no larvae from this last family were obtained in recovery).

Native Bush Habitat

This environment had recorded daily average maximum and minimum temperatures were 18.0 °C (64.4 °F) and 13.0 °C (55.4 °F), respectively. The average rainfall in this habitat was recorded at 0.4 mm/day. After the bloat stage, which lasted until day seven after death, post-active decay began around day 14. In this habitat, the *H. rostrata*, adult Phoridae, Sylvicola larvae and adult were the predominant species remaining on the body during the pre-skeletonization stages.

In Literature

Throughout its history the study of forensic entomology has not remained an esoteric science reserved only for entomologists and forensic scientists. Early twentieth-century popular scientific literature began to pique a broader interest in entomology. The very popular ten-volume book series, Alfred Brehem's *Thierleben* (Life of Animals, 1876–1879) expounded on many zoological topics, including arthropods. The accessible writing style of French entomologist Jean-Henri Fabre was also instrumental in the popularization of entomology. His collection of writings *Souvenirs Entomologique*, written during the last half of the 19th century, is especially useful because of the meticulous attention to detail to the observed insects' behaviors and life cycles.

The real impetus behind the modern cultural fascination with solving crime using entomological evidence can be traced back to the works *Faune des Tombeaux* (Fauna of the Tombs, 1887) and *Les Faunes des Cadavres* (Fauna of Corpses, 1894) by French veterinarian and entomologist Jean Pierre Mégnin. These works made the concept of the process of insect ecological succession on a corpse understandable and interesting to an ordinary reader in a way that no other previous scientific work had done. It was after the publication of Mégnin's work that the studies of forensic science and entomology became an established part of Western popular culture, which in turn inspired other scientists to continue and expand upon his research.

Economic Entomology

Economic entomology is a field of entomology, which involves the study of insects that benefit or harm humans, domestic animals, and crops. Insects that cause losses are termed as pests. Some species can cause indirect damage by spreading diseases and these are termed as vectors. Those that are beneficial include those reared for food such as honey, substances such as lac or pigments and for their role in pollinating crops and controlling pests.

History

In the 18th century many works were published on agriculture. Many contained accounts of pest insects. In France Claude Sionnest (1749–1820) was a notable figure.

19th Century

The most able exponent of this subject in Great Britain was John Curtis, whose treatise *Farm Insects*, published in 1860, was once the standard British work dealing with the insect pests of corn, roots, grass and stored corn. The most important works dealing with fruit and other pests were by Saunders, Joseph Albert Lintner, Charles Valentine Riley, Mark Vernon Slingerland and others in America and Canada. In Europe the earliest works were by Ernst Ludwig Taschenberg, Sven Lampa (1839–1914), Enzio Reuter (1867–1951) and Vincenze Kollar. Charles French (1842–1933), Walter Wilson Froggatt (1858–1937) and Henry Tryon (1856–1943) pioneered in Australia. It was not until the last quarter of the 19th century that any real advance was made in the study of economic entomology. Among the early writings, besides the book of Curtis, there was also a publication by Pohl and Kollar, entitled *Insects Injurious to Gardeners, Foresters and Farmers*, published in 1837, and Taschenberg's *Praktische Insecktenkunde*. During the 19th century Italian entomologists made significant progress in controlling diseases of the Silk moth which supported the silk industry, in the control of agricultural pests and in stored product entomology. Significant figures were: Agostino Bassi (1773–1856), Camillo Róndani (1808–1879), Adolfo Targioni Tozzetti (1823–1902), Pietro Stefanelli (1835, 1919), Camillo Acqua (1863–1936) Antonio Berlese (1863–1927), Gustavo Leonardi(1869–1918) and Enrico Verson (1845–1927). In France Etienne Laurent Joseph Hippolyte Boyer de Fonscolombe, Charles Jean-Baptiste Amyot, Émile Blanchard, Valéry Mayet and Claude Charles Goureau were early workers, as was Jean Victoire Audouin, the author of *Histoire des insectes nuisibles à la vigne et particulièrement de la Pyrale*, Philippe Alexandre Jules Künckel d'Herculais and Jean-Étienne Girard. American literature began as far back as 1788, when a report on the Hessian fly was issued by Sir Joseph Banks; in 1817 Thomas Say began his writings; while in 1856 Asa Fitch started his report on *Noxious Insects of New York*. Also in America, Matthew Cooke wrote *Treatise on the Insects Injurious to Fruit and Fruit Trees of the State of California, and Remedies Recommended for Their Extermination*, published in 1881. The Englishman Frederick Vincent Theobald wrote A textbook of agricultural zoology in 1890. It became a standard text worldwide. Notable foresters were Herman von Nördlinger (1818–1897) and Julius Theodor Christian Ratzeburg (1801–1871)

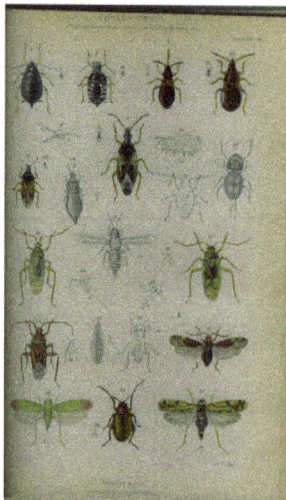

"Insects infesting potato crops": a plate from John Curtis's *Farm Insects*, 1860

20th Century

Among the most important reports early in the 20th century were those of Charles Valentine Riley, published by the U.S. Department of Agriculture, extending from 1878 to his death, in which is embodied an enormous amount of valuable material. At his death the work fell to Professor Leland Ossian Howard, in the form of *Bulletin of the U.S. Department of Agriculture*. The chief writings of J. A. Lintner extend from 1882 to 1898, in yearly parts, under the title of *Reports on the Injurious Insects of the State of New York*. Another significant contributor to the entomological literature of the United States was Charles W. Woodworth. The Florida entomologist Wilmon Newell was a pioneer of pest control as was Clarence Preston Gillette. In India Thomas Bainbrigge Fletcher, who succeeded Harold Maxwell-Lefroy and Lionel de Nicéville as the first Imperial Entomologist, wrote *Some South Indian insects and other animals of importance considered especially from an economic point of view*, an influential work in the subcontinent. In France Alfred Balachowsky was a key figure. In the last quarter of the 20th century new techniques were pioneered and new theories developed, for instance Integrated Pest Management by Ray F. Smith.

Harmful Insects

Insects considered *pests* of some sort occur among all major living orders with the exception of Ephemeroptera (mayflies), Odonata, Plecoptera (stoneflies), Embioptera (webspinners), Trichoptera (caddisflies), Neuroptera (in the broad sense), and Mecoptera (also, the tiny groups Zoraptera, Grylloblattodea, and Mantophasmatodea). Conversely, of course, essentially all insect orders primarily have members which are beneficial, in some respects, with the exception of Phthiraptera (lice), Siphonaptera (fleas), and Strepsiptera, the three orders whose members are exclusively parasitic.

Insects are considered as pests for a variety of reasons including their

- direct damage by feeding on crop plants in the field or by infesting stored products

- indirect damage by spreading viral diseases of crop plants (especially by sucking insects such as leafhoppers)

- spreading disease among humans and livestock

- annoyance to humans

Examples

- The Phylloxera plague

- Migratory locust

- Colorado potato beetle

- Boll weevil

- Japanese beetle

- Aphids

- Mosquitoes

- Cockroach

- Western corn rootworm

- Some fly species

The phylloxera, a true gourmet, finds out the best vineyards and attaches itself to the best wines
Cartoon from Punch, 6 September 1890)

In the past entomologists working on pest insects attempted to *eradicate* species. This has rarely worked except in islands or controlled environments and raises ethical issues. Over time the language has changed to terms like *control* and *management*. The indiscriminate use of toxic and persistent chemicals and the resurgence of pests in the history of cotton growing in the US has been particularly well studied.

Beneficial Insects

Honey is perhaps the most economically valuable product from insects. Apiculture is a commercial enterprise in most parts of the world and many forest tribes have been dependent on honey as a major source of nutrition. Honeybees can also act as pollinators of crop species. Many predators and parasitoid insects are encouraged and augmented in modern agriculture.

Boll Weevil Monument, erected by the citizens of Enterprise, Alabama to honour the pest that ended their dependence on cotton, a poverty crop.

Silk is extracted from both reared caterpillars as well as from the wild (producing wild silk). Sericulture deals with the techniques for efficient silkworm rearing and silk production. Although new fabric materials have substituted silk in many applications, it continues to be the material of choice for surgical sutures.

Lac was once extracted from scale insects but is now replaced by synthetic substitutes. The dye extracted from cochineal insects was similarly replaced by technological advances.

The idea of insects as human food, entomophagy, has been proposed as a solution to meet the growing demand for food, but has not gained widespread acceptance.

References

- Deborah Gordon (2010). Ant Encounters Interaction Networks and Colony Behavior. New Jersey: Princeton University Press. p. 143. ISBN 978-0691138794.

- Sleigh, Charlotte (2007) Six legs better : a cultural history of myrmecology. The Johns Hopkins University Press. ISBN 0-8018-8445-4

- Goddard, J. 2007. Physician's Guide to Arthropods of Medical Importance, Fifth Edition.Boca Raton, FL, CRC Press, ISBN 978-0-8493-8539-1 ISBN 0-8493-8539-3

- Service, M. 2008. Medical Entomology for Students 4th Edition Cambridge University Press. ISBN 978-0-521-70928-6

- R.H. van Gulik (2004) [1956]. T'and-Yin-Pi-Shih: Parallel cases from under the pear-tree (reprint ed.). Gibson Press. p. 18. ISBN 0-88355-908-0.

- Rutsch, Poncie (22 January 2015). "Finding Crime Clues In What Insects Had For Dinner". NPR. Retrieved 22 June 2015.

3

Insect and its Classification

Insects are the most diverse form of animals and represent more than a million species. Insects can be found in every habitat, although only a small number is found in the oceans. This chapter lists insects such as, beetle, hemiptera, termite, lepidopter, bee, ant, grasshopper and wasp. The chapter serves as a source to understand the major categories related to insects.

Insect

Insects are a class of invertebrates within the arthropod phylum that have a chitinous exoskeleton, a three-part body (head, thorax and abdomen), three pairs of jointed legs, compound eyes and one pair of antennae. They are the most diverse group of animals on the planet, including more than a million described species and representing more than half of all known living organisms. The number of extant species is estimated at between six and ten million, and potentially represent over 90% of the differing animal life forms on Earth. Insects may be found in nearly all environments, although only a small number of species reside in the oceans, a habitat dominated by another arthropod group, crustaceans.

The life cycles of insects vary but most hatch from eggs. Insect growth is constrained by the inelastic exoskeleton and development involves a series of molts. The immature stages can differ from the adults in structure, habit and habitat, and can include a passive pupal stage in those groups that undergo 4-stage metamorphosis. Insects that undergo 3-stage metamorphosis lack a pupal stage and adults develop through a series of nymphal stages. The higher level relationship of the Hexapoda is unclear. Fossilized insects of enormous size have been found from the Paleozoic Era, including giant dragonflies with wingspans of 55 to 70 cm (22–28 in). The most diverse insect groups appear to have coevolved with flowering plants.

Adult insects typically move about by walking, flying or sometimes swimming. As it allows for rapid yet stable movement, many insects adopt a tripedal gait in which they walk with their legs touching the ground in alternating triangles. Insects are the only invertebrates to have evolved flight. Many insects spend at least part of their lives under water, with larval adaptations that include gills, and some adult insects are aquatic and have adaptations for swimming. Some species, such as water striders, are capable of walking on the surface of water. Insects are mostly solitary, but some, such as certain bees, ants and termites, are social and live in large, well-organized colonies. Some insects, such as earwigs, show maternal care, guarding their eggs and young. Insects can communicate with each other in a variety of ways. Male moths can sense the pheromones of female moths over great distances. Other species communicate with sounds: crickets stridulate, or rub their wings together, to attract a mate and repel other males. Lampyridae in the beetle order communicate with light.

Humans regard certain insects as pests, and attempt to control them using insecticides and a host of other techniques. Some insects damage crops by feeding on sap, leaves or fruits. A few parasitic species are pathogenic. Some insects perform complex ecological roles; blow-flies, for example, help consume carrion but also spread diseases. Insect pollinators are essential to the life-cycle of many flowering plant species on which most organisms, including humans, are at least partly dependent; without them, the terrestrial portion of the biosphere (including humans) would be devastated. Many other insects are considered ecologically beneficial as predators and a few provide direct economic benefit. Silkworms and bees have been used extensively by humans for the production of silk and honey, respectively. In some cultures, people eat the larvae or adults of certain insects.

Etymology

The word "insect" comes from the Latin word *insectum*, meaning "with a notched or divided body", or literally "cut into", from the neuter singular perfect passive participle of *insectare*, "to cut into, to cut up", from *in-* "into" and *secare* "to cut"; because insects appear "cut into" three sections. "Insect" first appears documented in English in 1601 in Holland's translation of Pliny. Translations of Aristotle's term also form the usual word for "insect" in Welsh (trychfil, from *trychu* "to cut" and *mil*, "animal").

Phylogeny and Evolution

The evolutionary relationship of insects to other animal groups remains unclear.

Although traditionally grouped with millipedes and centipedes—possibly on the basis of convergent adaptations to terrestrialisation—evidence has emerged favoring closer evolutionary ties with crustaceans. In the Pancrustacea theory, insects, together with Entognatha, Remipedia, and Cephalocarida, make up a natural clade labeled Miracrustacea.

A report in November 2014 unambiguously places the insects in one clade, with the crustaceans and myriapods, as the nearest sister clades. This study resolved insect phylogeny of all extant insect orders, and provides "a robust phylogenetic backbone tree and reliable time estimates of insect evolution."

Evolution has produced enormous variety in insects. Pictured are some of the possible shapes of antennae.

Other terrestrial arthropods, such as centipedes, millipedes, scorpions, and spiders, are sometimes confused with insects since their body plans can appear similar, sharing (as do all arthropods) a jointed exoskeleton. However, upon closer examination, their features differ significantly; most noticeably, they do not have the six-legged characteristic of adult insects.

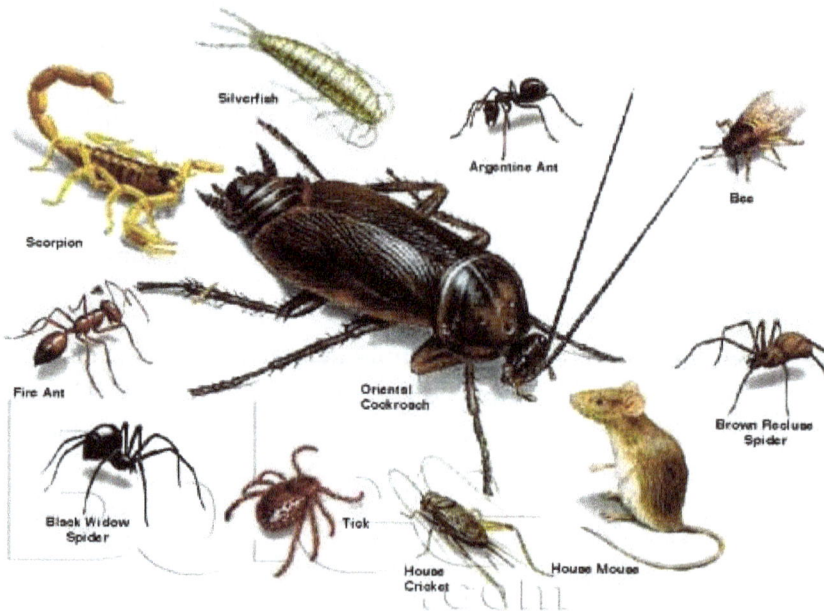

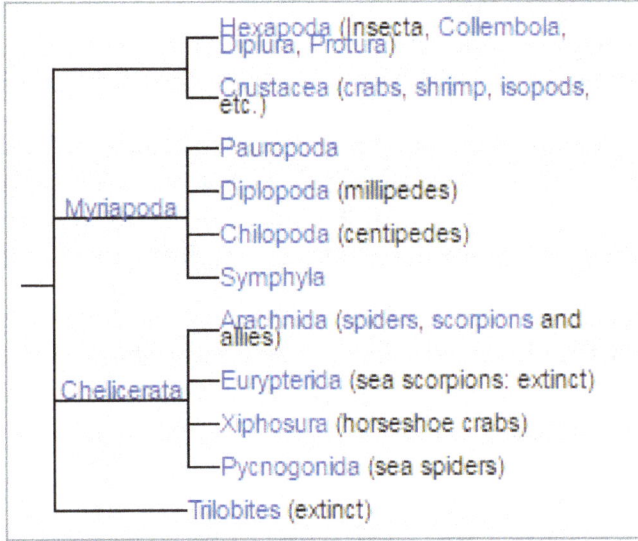

A phylogenetic tree of the arthropods and related groups

The higher-level phylogeny of the arthropods continues to be a matter of debate and research. In 2008, researchers at Tufts University uncovered what they believe is the world's oldest known full-body impression of a primitive flying insect, a 300 million-year-old specimen from the Carboniferous period. The oldest definitive insect fossil is the Devonian *Rhyniognatha hirsti*, from the 396-million-year-old Rhynie chert. It may have superficially resembled a modern-day silverfish insect. This species already possessed dicondylic mandibles (two articulations in the mandible), a feature associated with winged insects, suggesting that wings may already have evolved at this time. Thus, the first insects probably appeared earlier, in the Silurian period.

Four super radiations of insects have occurred: beetles (evolved about 300 million years ago), flies (evolved about 250 million years ago), and moths and wasps (evolved about 150 million years ago). These four groups account for the majority of described species. The flies and moths along with the fleas evolved from the Mecoptera.

The origins of insect flight remain obscure, since the earliest winged insects currently known appear to have been capable fliers. Some extinct insects had an additional pair of winglets attaching to the first segment of the thorax, for a total of three pairs. As of 2009, no evidence suggests the insects were a particularly successful group of animals before they evolved to have wings.

Late Carboniferous and Early Permian insect orders include both extant groups, their stem groups, and a number of Paleozoic groups, now extinct. During this era, some giant dragonfly-like forms reached wingspans of 55 to 70 cm (22 to 28 in), making them far larger than any living insect. This gigantism may have been due to higher atmospheric oxygen levels that allowed increased respiratory efficiency relative to today. The lack of flying vertebrates could have been another factor. Most extinct orders of insects developed during the Permian period that began around 270 million years ago. Many of the early groups became extinct during the Permian-Triassic extinction event, the largest mass extinction in the history of the Earth, around 252 million years ago.

The remarkably successful Hymenoptera appeared as long as 146 million years ago in the Cretaceous period, but achieved their wide diversity more recently in the Cenozoic era, which began 66 million years ago. A number of highly successful insect groups evolved in conjunction with flowering plants, a powerful illustration of coevolution.

Many modern insect genera developed during the Cenozoic. Insects from this period on are often found preserved in amber, often in perfect condition. The body plan, or morphology, of such specimens is thus easily compared with modern species. The study of fossilized insects is called paleoentomology.

Evolutionary Relationships

Insects are prey for a variety of organisms, including terrestrial vertebrates. The earliest vertebrates on land existed 400 million years ago and were large amphibious piscivores. Through gradual evolutionary change, insectivory was the next diet type to evolve.

Insects were among the earliest terrestrial herbivores and acted as major selection agents on plants. Plants evolved chemical defenses against this herbivory and the insects, in turn, evolved mechanisms to deal with plant toxins. Many insects make use of these toxins to protect themselves from their predators. Such insects often advertise their toxicity using warning colors. This successful evolutionary pattern has also been used by mimics. Over time, this has led to complex groups of coevolved species. Conversely, some interactions between plants and insects, like pollination, are beneficial to both organisms. Coevolution has led to the development of very specific mutualisms in such systems.

Taxonomy

Traditional morphology-based or appearance-based systematics have usually given the Hexapoda the rank of superclass,and identified four groups within it: insects (Ectognatha), springtails (Collembola), Protura, and Diplura, the latter three being grouped together as the Entognatha on the basis of internalized mouth parts. Supraordinal relationships have undergone numerous changes with the advent of methods based on evolutionary history and genetic data. A recent theory is that the Hexapoda are polyphyletic (where the last common ancestor was not a member of the group), with the entognath classes having separate evolutionary histories from the Insecta. Many of the traditional appearance-based taxa have been shown to be paraphyletic, so rather than using ranks like subclass, superorder, and infraorder, it has proved better to use monophyletic groupings (in which the last common ancestor is a member of the group). The following represents the best-supported monophyletic groupings for the Insecta.

Insects can be divided into two groups historically treated as subclasses: wingless insects, known as Apterygota, and winged insects, known as Pterygota. The Apterygota consist of the primitively wingless order of the silverfish (Thysanura). Archaeognatha make up the Monocondylia based on the shape of their mandibles, while Thysanura and Pterygota are grouped together as Dicondylia. The Thysanura themselves possibly are not monophyletic, with the family Lepidotrichidae being a sister group to the Dicondylia (Pterygota and the remaining Thysanura).

Paleoptera and Neoptera are the winged orders of insects differentiated by the presence of hardened body parts called sclerites, and in the Neoptera, muscles that allow their wings to fold flatly over the abdomen. Neoptera can further be divided into incomplete metamorphosis-based (Polyneoptera and Paraneoptera) and complete metamorphosis-based groups. It has proved difficult to clarify the relationships between the orders in Polyneoptera because of constant new findings calling for revision of the taxa. For example, the Paraneoptera have turned out to be more closely related to the Endopterygota than to the rest of the Exopterygota. The recent molecular finding that the traditional louse orders Mallophaga and Anoplura are derived from within Psocoptera has led to the new taxon Psocodea. Phasmatodea and Embiidina have been suggested to form the Eukinolabia. Mantodea, Blattodea, and Isoptera are thought to form a monophyletic group termed Dictyoptera.

The Exopterygota likely are paraphyletic in regard to the Endopterygota. Matters that have incurred controversy include Strepsiptera and Diptera grouped together as Halteria based on a reduction of one of the wing pairs – a position not well-supported in the entomological community. The Neuropterida are often lumped or split on the whims of the taxonomist. Fleas are now thought to be closely related to boreid mecopterans. Many questions remain in the basal relationships amongst endopterygote orders, particularly the Hymenoptera.

The study of the classification or taxonomy of any insect is called systematic entomology. If one works with a more specific order or even a family, the term may also be made specific to that order or family, for example systematic dipterology.

Diversity

Though the true dimensions of species diversity remain uncertain, estimates range from 2.6–7.8 million species with a mean of 5.5 million. This probably represents less than 20% of all species on Earth, and with only about 20,000 new species of all organisms being described each year, most species likely will remain undescribed for many years unless species descriptions increase in rate. About 850,000–1,000,000 of all described species are insects. Of the 24 orders of insects, four dominate in terms of numbers of described species, with at least 3 million species included in Coleoptera, Diptera, Hymenoptera and Lepidoptera. A recent study estimated the number of beetles at 0.9–2.1 million with a mean of 1.5 million.

Comparison of the estimated number of species in the four most speciose insect orders			
	Described species	**Average description rate (species per year)**	**Publication effort**
Coleoptera	300,000–400,000	2308	0.01
Lepidoptera	110,000–120,000	642	0.03
Diptera	90,000–150,000	1048	0.04
Hymenoptera	100,000–125,000	1196	0.02

Morphology and Physiology

External

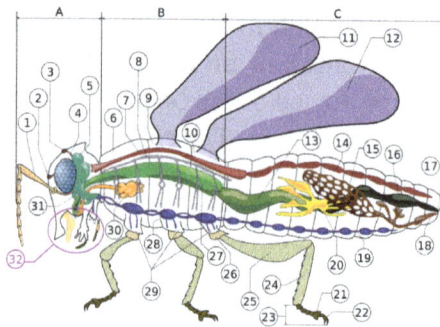

Insect morphology
A- Head B- Thorax C- Abdomen

1. antenna
2. ocelli (lower)
3. ocelli (upper)
4. compound eye
5. brain (cerebral ganglia)
6. prothorax
7. dorsal blood vessel
8. tracheal tubes (trunk with spiracle)
9. mesothorax
10. metathorax
11. forewing
12. hindwing
13. mid-gut (stomach)
14. dorsal tube (Heart)
15. ovary
16. hind-gut (intestine, rectum & anus)
17. anus
18. oviduct
19. nerve chord (abdominal ganglia)
20. Malpighian tubes

21. tarsal pads
22. claws
23. tarsus
24. tibia
25. femur
26. trochanter
27. fore-gut (crop, gizzard)
28. thoracic ganglion
29. coxa
30. salivary gland
31. subesophageal ganglion
32. mouthparts

Insects have segmented bodies supported by exoskeletons, the hard outer covering made mostly of chitin. The segments of the body are organized into three distinctive but interconnected units, or tagmata: a head, a thorax and an abdomen. The head supports a pair of sensory antennae, a pair of compound eyes, and, if present, one to three simple eyes (or ocelli) and three sets of variously modified appendages that form the mouthparts. The thorax has six segmented legs—one pair each for the prothorax, mesothorax and the metathorax segments making up the thorax—and, none, two or four wings. The abdomen consists of eleven segments, though in a few species of insects, these segments may be fused together or reduced in size. The abdomen also contains most of the digestive, respiratory, excretory and reproductive internal structures. Considerable variation and many adaptations in the body parts of insects occur, especially wings, legs, antenna and mouthparts.

Segmentation

The head is enclosed in a hard, heavily sclerotized, unsegmented, exoskeletal head capsule, or epicranium, which contains most of the sensing organs, including the antennae, ocellus or eyes, and the mouthparts. Of all the insect orders, Orthoptera displays the most features found in other insects, including the sutures and sclerites. Here, the vertex, or the apex (dorsal region), is situated between the compound eyes for insects with a hypognathous and opisthognathous head. In prognathous insects, the vertex is not found between the compound eyes, but rather, where the ocelli are normally. This is because the primary axis of the head is rotated 90° to become parallel to the primary axis of the body. In some species, this region is modified and assumes a different name.

The thorax is a tagma composed of three sections, the prothorax, mesothorax and the metathorax. The anterior segment, closest to the head, is the prothorax, with the major features being the first pair of legs and the pronotum. The middle segment is the mesothorax, with the major features being the second pair of legs and the anterior wings. The third and most posterior segment, abutting the abdomen, is the metathorax, which features the third pair of legs and the posterior wings. Each segment is dilineated by an intersegmental suture. Each segment has four basic regions. The dorsal surface is called the tergum (or *notum*) to distinguish it from the abdominal terga. The two lateral regions are called the pleura (singular: pleuron) and the ventral aspect is called the sternum. In turn, the notum of the prothorax is called the pronotum, the notum for the mesothorax is called the mesonotum and the notum for the metathorax is called the metanotum. Continuing with this logic, the mesopleura and metapleura, as well as the mesosternum and metasternum, are used.

The abdomen is the largest tagma of the insect, which typically consists of 11–12 segments and is less strongly sclerotized than the head or thorax. Each segment of the abdomen is represented by a sclerotized tergum and sternum. Terga are separated from each other and from the adjacent sterna or pleura by membranes. Spiracles are located in the pleural area. Variation of this ground plan includes the fusion of terga or terga and sterna to form continuous dorsal or ventral shields or a conical tube. Some insects bear a sclerite in the pleural area called a laterotergite. Ventral sclerites are sometimes called laterosternites. During the embryonic stage of many insects and the postembryonic stage of primitive insects, 11 abdominal segments are present. In modern insects there is a tendency toward reduction in the number of the abdominal segments, but the primitive number of 11 is maintained during embryogenesis. Variation in abdominal segment number is considerable. If the Apterygota are considered to be indicative of the ground plan for pterygotes, confusion reigns: adult Protura have 12 segments, Collembola have 6. The orthopteran family Acrididae has 11 segments, and a fossil specimen of Zoraptera has a 10-segmented abdomen.

Exoskeleton

The insect outer skeleton, the cuticle, is made up of two layers: the epicuticle, which is a thin and waxy water resistant outer layer and contains no chitin, and a lower layer called the procuticle. The procuticle is chitinous and much thicker than the epicuticle and has two layers: an outer layer known as the exocuticle and an inner layer known as the endocuticle. The tough and flexible endocuticle is built from numerous layers of fibrous chitin and proteins, criss-crossing each other in a sandwich pattern, while the exocuticle is rigid and hardened. The exocuticle is greatly reduced in many soft-bodied insects (e.g., caterpillars), especially during their larval stages.

Insects are the only invertebrates to have developed active flight capability, and this has played an important role in their success. Their muscles are able to contract multiple times for each single nerve impulse, allowing the wings to beat faster than would ordinarily be possible. Having their muscles attached to their exoskeletons is more efficient and allows more muscle connections; crustaceans also use the same method, though all spiders use hydraulic pressure to extend their legs, a system inherited from their pre-arthropod ancestors. Unlike insects, though, most aquatic crustaceans are biomineralized with calcium carbonate extracted from the water.

Internal

Nervous System

The nervous system of an insect can be divided into a brain and a ventral nerve cord. The head capsule is made up of six fused segments, each with either a pair of ganglia, or a cluster of nerve cells outside of the brain. The first three pairs of ganglia are fused into the brain, while the three following pairs are fused into a structure of three pairs of ganglia under the insect's esophagus, called the subesophageal ganglion.

The thoracic segments have one ganglion on each side, which are connected into a pair, one pair per segment. This arrangement is also seen in the abdomen but only in the first eight segments. Many species of insects have reduced numbers of ganglia due to fusion or reduction. Some cockroaches have just six ganglia in the abdomen, whereas the wasp *Vespa crabro* has only two in the thorax and three in the abdomen. Some insects, like the house fly *Musca domestica*, have all the body ganglia fused into a single large thoracic ganglion.

At least a few insects have nociceptors, cells that detect and transmit signals responsible for the sensation of pain. This was discovered in 2003 by studying the variation in reactions of larvae of the common fruitfly Drosophila to the touch of a heated probe and an unheated one. The larvae reacted to the touch of the heated probe with a stereotypical rolling behavior that was not exhibited when the larvae were touched by the unheated probe. Although nociception has been demonstrated in insects, there is no consensus that insects feel pain consciously

Insects are capable of learning.

Digestive System

An insect uses its digestive system to extract nutrients and other substances from the food it consumes. Most of this food is ingested in the form of macromolecules and other complex substances like proteins, polysaccharides, fats and nucleic acids. These macromolecules must be broken down by catabolic reactions into smaller molecules like amino acids and simple sugars before being used by cells of the body for energy, growth, or reproduction. This break-down process is known as digestion.

The main structure of an insect's digestive system is a long enclosed tube called the alimentary canal, which runs lengthwise through the body. The alimentary canal directs food unidirectionally from the mouth to the anus. It has three sections, each of which performs a different process of digestion. In addition to the alimentary canal, insects also have paired salivary glands and salivary reservoirs. These structures usually reside in the thorax, adjacent to the foregut.

The salivary glands (element 30 in numbered diagram) in an insect's mouth produce saliva. The salivary ducts lead from the glands to the reservoirs and then forward through the head to an opening called the salivarium, located behind the hypopharynx. By moving its mouthparts (element 32 in numbered diagram) the insect can mix its food with saliva. The mixture of saliva and food then travels through the salivary tubes into the mouth, where it begins to break down. Some insects, like flies, have extra-oral digestion. Insects using extra-oral digestion expel digestive enzymes onto their food to break it down. This strategy allows insects to extract a significant proportion of the available nutrients from the food source. The gut is where almost all of insects' digestion takes place. It can be divided into the foregut, midgut and hindgut.

Foregut

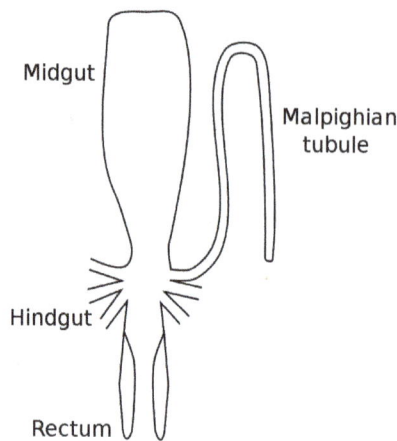

Stylized diagram of insect digestive tract showing malpighian tubule, from an insect of the order Orthoptera

The first section of the alimentary canal is the foregut (element 27 in numbered diagram), or stomodaeum. The foregut is lined with a cuticular lining made of chitin and proteins as protection from tough food. The foregut includes the buccal cavity (mouth), pharynx, esophagus and crop and proventriculus (any part may be highly modified) which both store food and signify when to continue passing onward to the midgut.

Digestion starts in buccal cavity (mouth) as partially chewed food is broken down by saliva from the salivary glands. As the salivary glands produce fluid and carbohydrate-digesting enzymes (mostly amylases), strong muscles in the pharynx pump fluid into the buccal cavity, lubricating the food like the salivarium does, and helping blood feeders, and xylem and phloem feeders.

From there, the pharynx passes food to the esophagus, which could be just a simple tube passing it on to the crop and proventriculus, and then onward to the midgut, as in most insects. Alternately, the foregut may expand into a very enlarged crop and proventriculus, or the crop could just be a diverticulum, or fluid-filled structure, as in some Diptera species.

Bumblebee defecating. Note the contraction of the abdomen to provide internal pressure

Midgut

Once food leaves the crop, it passes to the midgut (element 13 in numbered diagram), also known as the mesenteron, where the majority of digestion takes place. Microscopic projections from the midgut wall, called microvilli, increase the surface area of the wall and allow more nutrients to be absorbed; they tend to be close to the origin of the midgut. In some insects, the role of the microvilli and where they are located may vary. For example, specialized microvilli producing digestive enzymes may more likely be near the end of the midgut, and absorption near the origin or beginning of the midgut.

Hindgut

In the hindgut (element 16 in numbered diagram), or proctodaeum, undigested food particles are joined by uric acid to form fecal pellets. The rectum absorbs 90% of the water in these fecal pellets, and the dry pellet is then eliminated through the anus (element 17), completing the process of digestion. The uric acid is formed using hemolymph waste products diffused from the Malpighian tubules (element 20). It is then emptied directly into the alimentary canal, at the junction between the midgut and hindgut. The number of Malpighian tubules possessed by a given insect varies between species, ranging from only two tubules in some insects to over 100 tubules in others.

Reproductive System

The reproductive system of female insects consist of a pair of ovaries, accessory glands, one or more spermathecae, and ducts connecting these parts. The ovaries are made up of a number of egg tubes, called ovarioles, which vary in size and number by species. The number of eggs that the insect is able to make vary by the number of ovarioles with the rate that eggs can be develop being also influenced by ovariole design. Female insects are able make eggs, receive and store sperm, manipulate sperm from different males, and lay eggs. Accessory glands or glandular parts of the oviducts produce a variety of substances for sperm maintenance, transport and fertilization, as well as for protection of eggs. They can produce glue and protective substances for coating eggs or tough coverings for a batch of eggs called oothecae. Spermathecae are tubes or sacs in which sperm can be stored between the time of mating and the time an egg is fertilized.

For males, the reproductive system is the testis, suspended in the body cavity by tracheae and the fat body. Most male insects have a pair of testes, inside of which are sperm tubes or follicles that are enclosed within a membranous sac. The follicles connect to the vas deferens by the vas efferens, and the two tubular vasa deferentia connect to a median ejaculatory duct that leads to the outside. A portion of the vas deferens is often enlarged to form the seminal vesicle, which stores the sperm before they are discharged into the female. The seminal vesicles have glandular linings that secrete nutrients for nourishment and maintenance of the sperm. The ejaculatory duct is derived from an invagination of the epidermal cells during development and, as a result, has a cuticular lining. The terminal portion of the ejaculatory duct may be sclerotized to form the intromittent organ, the aedeagus. The remainder of the male reproductive system is derived from embryonic mesoderm, except for the germ cells, or spermatogonia, which descend from the primordial pole cells very early during embryogenesis.

Respiratory System

The tube-like heart (green) of the mosquito *Anopheles gambiae* extends horizontally across the body, interlinked with the diamond-shaped wing muscles (also green) and surrounded by pericardial cells (red). Blue depicts cell nuclei.

Insect respiration is accomplished without lungs. Instead, the insect respiratory system uses a system of internal tubes and sacs through which gases either diffuse or are actively pumped, delivering oxygen directly to tissues that need it via their trachea (element 8 in numbered diagram). Since oxygen is delivered directly, the circulatory system is not used to carry oxygen, and is therefore greatly reduced. The insect circulatory system has no veins or arteries, and instead consists of little more than a single, perforated dorsal tube which pulses peristaltically. Toward the thorax, the dorsal tube (element 14) divides into chambers and acts like the insect's heart. The opposite end of the dorsal tube is like the aorta of the insect circulating the hemolymph, arthropods' fluid analog of blood, inside the body cavity. Air is taken in through openings on the sides of the abdomen called spiracles.

The respiratory system is an important factor that limits the size of insects. As insects get bigger, this type of oxygen transport gets less efficient and thus the heaviest insect currently weighs less than 100 g. However, with increased atmospheric oxygen levels, as happened in the late Paleozoic, larger insects were possible, such as dragonflies with wingspans of more than two feet.

There are many different patterns of gas exchange demonstrated by different groups of insects. Gas exchange patterns in insects can range from continuous and diffusive ventilation, to discontinuous gas exchange. During continuous gas exchange, oxygen is taken in and carbon dioxide is released in a continuous cycle. In discontinuous gas exchange, however, the insect takes in oxygen while it is active and small amounts of carbon dioxide are released when the insect is at rest. Diffusive ventilation is simply a form of continuous gas exchange that occurs by diffusion rather than physically taking in the oxygen. Some species of insect that are submerged also have adaptations to aid in respiration. As larvae, many insects have gills that can extract oxygen dissolved in water, while others need to rise to the water surface to replenish air supplies which may be held or trapped in special structures.

Circulatory System

The insect circulatory system utilizes hemolymph, a tissue analogous to blood that circulates in the interior of the insect body, while remaining in direct contact with the animal's tissues. It is composed of plasma in which hemocytes are suspended. In addition to hemocytes, the plasma also contains many chemicals. It is also the major tissue type of the open circulatory system of arthropods, characteristic of spiders, crustaceans and insects.

Reproduction and Development

A pair of Simosyrphus grandicornis hoverflies mating in flight.

A pair of grasshoppers mating.

The majority of insects hatch from eggs. The fertilization and development takes place inside the egg, enclosed by a shell (chorion) that consists of maternal tissue. In contrast to eggs of other arthropods, most insect eggs are drought resistant. This is because inside the chorion two additional membranes develop from embryonic tissue, the amnion and the serosa. This serosa secretes a cuticle rich in chitin that protects the embryo against desiccation. In Schizophora however the serosa does not develop, but these flies lay their eggs in damp places, such as rotting matter. Some species of insects, like the cockroach *Blaptica dubia*, as well as juvenile aphids and tsetse flies, are ovoviviparous. The eggs of ovoviviparous animals develop entirely inside the female, and then hatch immediately upon being laid. Some other species, such as those in the genus of cockroaches known as *Diploptera*, are viviparous, and thus gestate inside the mother and are born alive. Some insects, like parasitic wasps, show polyembryony, where a single fertilized egg divides into many and in some cases thousands of separate embryos. Insects may be *univoltine, bivoltine* or *multivoltine*, i.e. they may have one, two or many broods (generations) in a year.

The different forms of the male (top) and female (bottom) tussock moth *Orgyia recens* is an example of sexual dimorphism in insects.

Other developmental and reproductive variations include haplodiploidy, polymorphism, paedomorphosis or peramorphosis, sexual dimorphism, parthenogenesis and more rarely hermaphroditism. In haplodiploidy, which is a type of sex-determination system, the offspring's sex is determined by the number of sets of chromosomes an individual receives. This system is typical in bees and wasps. Polymorphism is where a species may have different *morphs* or *forms*, as in the oblong winged katydid, which has four different varieties: green, pink and yellow or tan. Some insects may retain phenotypes that are normally only seen in juveniles; this is called paedomorphosis. In peramorphosis, an opposite sort of phenomenon, insects take on previously unseen traits after they have matured into adults. Many insects display sexual dimorphism, in which males and females have notably different appearances, such as the moth *Orgyia recens* as an exemplar of sexual dimorphism in insects.

Some insects use parthenogenesis, a process in which the female can reproduce and give birth without having the eggs fertilized by a male. Many aphids undergo a form of parthenogenesis, called cyclical parthenogenesis, in which they alternate between one or many generations of asexual and sexual reproduction. In summer, aphids are generally female and parthenogenetic; in the autumn, males may be produced for sexual reproduction. Other insects produced by parthenogenesis are bees, wasps and ants, in which they spawn males. However, overall, most individuals are female, which are produced by fertilization. The males are haploid and the females are diploid. More rarely, some insects display hermaphroditism, in which a given individual has both male and female reproductive organs.

Insect life-histories show adaptations to withstand cold and dry conditions. Some temperate region insects are capable of activity during winter, while some others migrate to a warmer climate or go into a state of torpor. Still other insects have evolved mechanisms of diapause that allow eggs or pupae to survive these conditions.

Metamorphosis

Metamorphosis in insects is the biological process of development all insects must undergo. There are two forms of metamorphosis: incomplete metamorphosis and complete metamorphosis.

Incomplete Metamorphosis

Hemimetabolous insects, those with incomplete metamorphosis, change gradually by undergoing a series of molts. An insect molts when it outgrows its exoskeleton, which does not stretch and would otherwise restrict the insect's growth. The molting process begins as the insect's epidermis secretes a new epicuticle inside the old one. After this new epicuticle is secreted, the epidermis releases a mixture of enzymes that digests the endocuticle and thus detaches the old cuticle. When this stage is complete, the insect makes its body swell by taking in a large quantity of water or air, which makes the old cuticle split along predefined weaknesses where the old exocuticle was thinnest.

Immature insects that go through incomplete metamorphosis are called nymphs or in the case of dragonflies and damselflies, also naiads. Nymphs are similar in form to the adult except for the presence of wings, which are not developed until adulthood. With each molt, nymphs grow larger and become more similar in appearance to adult insects.

This Southern Hawker dragonfly molts its exoskeleton several times during its life as a nymph; shown is the final molt to become a winged adult (eclosion).

Complete Metamorphosis

Gulf Fritillary Life Cycle

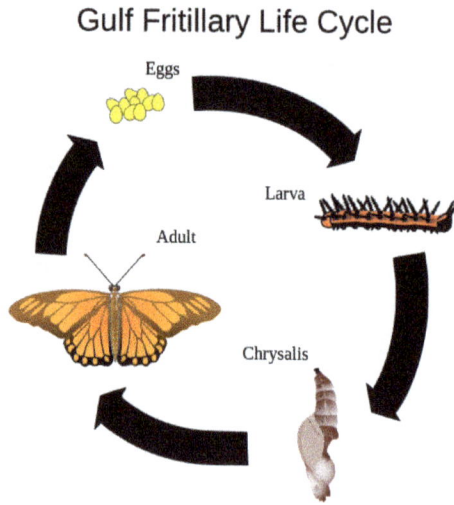

Gulf Fritillary life cycle, an example of holometabolism.

Holometabolism, or complete metamorphosis, is where the insect changes in four stages, an egg or embryo, a larva, a pupa and the adult or imago. In these species, an egg hatches to produce a larva, which is generally worm-like in form. This worm-like form can be one of several varieties: eruciform (caterpillar-like), scarabaeiform (grub-like), campodeiform (elongated, flattened and active), elateriform (wireworm-like) or vermiform (maggot-like). The larva grows and eventually becomes a pupa, a stage marked by reduced movement and often sealed within a cocoon. There are three types of pupae: obtect, exarate or coarctate. Obtect pupae are compact, with the legs and other appendages enclosed. Exarate pupae have their legs and other appendages free and extended. Coarctate pupae develop inside the larval skin. Insects undergo considerable change in form during the pupal stage, and emerge as adults. Butterflies are a well-known example of insects that undergo complete metamorphosis, although most insects use this life cycle. Some insects have evolved this system to hypermetamorphosis.

Some of the oldest and most successful insect groups, such Endopterygota, use a system of complete metamorphosis. Complete metamorphosis is unique to a group of certain insect orders including Diptera, Lepidoptera and Hymenoptera. This form of development is exclusive and not seen in any other arthropods.

Senses and Communication

Many insects possess very sensitive and, or specialized organs of perception. Some insects such as bees can perceive ultraviolet wavelengths, or detect polarized light, while the antennae of male moths can detect the pheromones of female moths over distances of many kilometers. The yellow paper wasp (*Polistes versicolor*) is known for its wagging movements as a form of communication within the colony; it can waggle with a frequency of 10.6±2.1 Hz (n=190). These wagging movements can signal the arrival of new material into the nest and aggression between workers can be used to stimulate others to increase foraging expeditions. There is a pronounced tendency for there to be a trade-off between visual acuity and chemical or tactile acuity, such that most insects with well-developed eyes have reduced or simple antennae, and vice versa. There are a variety of different mechanisms by which insects perceive sound, while the patterns are not universal, insects can generally hear sound if they can produce it. Different insect species can have varying hearing, though most insects can hear only a narrow range of frequencies related to the frequency of the sounds they can produce. Mosquitoes have been found to hear up to 2 kHz, and some grasshoppers can hear up to 50 kHz. Certain predatory and parasitic insects can detect the characteristic sounds made by their prey or hosts, respectively. For instance, some nocturnal moths can perceive the ultrasonic emissions of bats, which helps them avoid predation. Insects that feed on blood have special sensory structures that can detect infrared emissions, and use them to home in on their hosts.

Some insects display a rudimentary sense of numbers, such as the solitary wasps that prey upon a single species. The mother wasp lays her eggs in individual cells and provides each egg with a number of live caterpillars on which the young feed when hatched. Some species of wasp always provide five, others twelve, and others as high as twenty-four caterpillars per cell. The number of caterpillars is different among species, but always the same for each sex of larva. The male solitary wasp in the genus *Eumenes* is smaller than the female, so the mother of one species supplies him with only five caterpillars; the larger female receives ten caterpillars in her cell.

Light Production and Vision

Insects have compound eyes and two antennae.

A few insects, such as members of the families Poduridae and Onychiuridae (Collembola), Mycetophilidae (Diptera) and the beetle families Lampyridae, Phengodidae, Elateridae and Staphylinidae are bioluminescent. The most familiar group are the fireflies, beetles of the family Lampyridae. Some species are able to control this light generation to produce flashes. The function varies with some species using them to attract mates, while others use them to lure prey. Cave dwelling larvae of *Arachnocampa* (Mycetophilidae, Fungus gnats) glow to lure small flying insects into sticky strands of silk. Some fireflies of the genus *Photuris* mimic the flashing of female *Photinus* species to attract males of that species, which are then captured and devoured. The colors of emitted light vary from dull blue (*Orfelia fultoni*, Mycetophilidae) to the familiar greens and the rare reds (*Phrixothrix tiemanni*, Phengodidae).

Most insects, except some species of cave crickets, are able to perceive light and dark. Many species have acute vision capable of detecting minute movements. The eyes may include simple eyes or ocelli as well as compound eyes of varying sizes. Many species are able to detect light in the infrared, ultraviolet and the visible light wavelengths. Color vision has been demonstrated in many species and phylogenetic analysis suggests that UV-green-blue trichromacy existed from at least the Devonian period between 416 and 359 million years ago.

Sound Production and Hearing

Insects were the earliest organisms to produce and sense sounds. Insects make sounds mostly by mechanical action of appendages. In grasshoppers and crickets, this is achieved by stridulation. Cicadas make the loudest sounds among the insects by producing and amplifying sounds with special modifications to their body and musculature. The African cicada *Brevisana brevis* has been measured at 106.7 decibels at a distance of 50 cm (20 in). Some insects, such as the *Helicoverpa zea* moths, hawk moths and Hedylid butterflies, can hear ultrasound and take evasive action when they sense that they have been detected by bats. Some moths produce ultrasonic clicks that were once thought to have a role in jamming bat echolocation. The ultrasonic clicks were subsequently found to be produced mostly by unpalatable moths to warn bats, just as warning colorations are used against predators that hunt by sight. Some otherwise palatable moths have evolved to mimic these calls. More recently, the claim that some moths can jam bat sonar has been revisited. Ultrasonic recording and high-speed infrared videography of bat-moth interactions suggest the palatable tiger moth really does defend against attacking big brown bats using ultrasonic clicks that jam bat sonar.

Very low sounds are also produced in various species of Coleoptera, Hymenoptera, Lepidoptera, Mantodea and Neuroptera. These low sounds are simply the sounds made by the insect's movement. Through microscopic stridulatory structures located on the insect's muscles and joints, the normal sounds of the insect moving are amplified and can be used to warn or communicate with other insects. Most sound-making insects also have tympanal organs that can perceive airborne sounds. Some species in Hemiptera, such as the corixids (water boatmen), are known to communicate via underwater sounds. Most insects are also able to sense vibrations transmitted through surfaces.

Communication using surface-borne vibrational signals is more widespread among insects because of size constraints in producing air-borne sounds. Insects cannot effectively produce low-frequency sounds, and high-frequency sounds tend to disperse more in a dense environment (such as foliage), so insects living in such environments communicate primarily using substrate-borne vibrations. The mechanisms of production of vibrational signals are just as diverse as those for producing sound in insects.

Some species use vibrations for communicating within members of the same species, such as to attract mates as in the songs of the shield bug *Nezara viridula*. Vibrations can also be used to communicate between entirely different species; lycaenid (gossamer-winged butterfly) caterpillars which are myrmecophilous (living in a mutualistic association with ants) communicate with ants in this way. The Madagascar hissing cockroach has the ability to press air through its spiracles to make a hissing noise as a sign of aggression; the Death's-head Hawkmoth makes a squeaking noise by forcing air out of their pharynx when agitated, which may also reduce aggressive worker honey bee behavior when the two are in close proximity.

Chemical Communication

Chemical communications in animals rely on a variety of aspects including taste and smell. Chemoreception is the physiological response of a sense organ (i.e. taste or smell) to a chemical stimulus where the chemicals act as signals to regulate the state or activity of a cell. A semiochemical is a message-carrying chemical that is meant to attract, repel, and convey information. Types of semiochemicals include pheromones and kairomones. One example is the butterfly *Phengaris arion* which uses chemical signals as a form of mimicry to aid in predation.

In addition to the use of sound for communication, a wide range of insects have evolved chemical means for communication. These chemicals, termed semiochemicals, are often derived from plant metabolites include those meant to attract, repel and provide other kinds of information. Pheromones, a type of semiochemical, are used for attracting mates of the opposite sex, for aggregating conspecific individuals of both sexes, for deterring other individuals from approaching, to mark a trail, and to trigger aggression in nearby individuals. Allomonea benefit their producer by the effect they have upon the receiver. Kairomones benefit their receiver instead of their producer. Synomones benefit the producer and the receiver. While some chemicals are targeted at individuals of the same species, others are used for communication across species. The use of scents is especially well known to have developed in social insects.

Social Behavior

Social insects, such as termites, ants and many bees and wasps, are the most familiar species of eusocial animal. They live together in large well-organized colonies that may be so tightly integrated and genetically similar that the colonies of some species are sometimes considered superorganisms. It is sometimes argued that the various species of honey bee are the only invertebrates (and indeed one of the few non-human groups) to have evolved a system of abstract symbolic communication where a behavior is used to *represent* and convey specific information about something in the environment. In this communication system, called dance language, the angle at which a bee dances represents a direction relative to the sun, and the length of the dance represents the distance to be flown. Though perhaps not as advanced as honey bees, bumblebees also potentially have some social communication behaviors. *Bombus terrestris*, for example, exhibit a faster learning curve for visiting unfamiliar, yet rewarding flowers, when they can see a conspecific foraging on the same species.

A cathedral mound created by termites (Isoptera).

Only insects which live in nests or colonies demonstrate any true capacity for fine-scale spatial orientation or homing. This can allow an insect to return unerringly to a single hole a few millimeters in diameter among thousands of apparently identical holes clustered together, after a trip of up to several kilometers' distance. In a phenomenon known as philopatry, insects that hibernate have shown the ability to recall a specific location up to a year after last viewing the area of interest. A few insects seasonally migrate large distances between different geographic regions (e.g., the overwintering areas of the Monarch butterfly).

Care of Young

The eusocial insects build nest, guard eggs, and provide food for offspring full-time. Most insects, however, lead short lives as adults, and rarely interact with one another except to mate or compete for mates. A small number exhibit some form of parental care, where they will at least guard their eggs, and sometimes continue guarding their offspring until adulthood, and possibly even feeding them. Another simple form of parental care is to construct a nest (a burrow or an actual construction, either of which may be simple or complex), store provisions in it, and lay an egg upon those provisions. The adult does not contact the growing offspring, but it nonetheless does provide food. This sort of care is typical for most species of bees and various types of wasps.

Locomotion

Flight

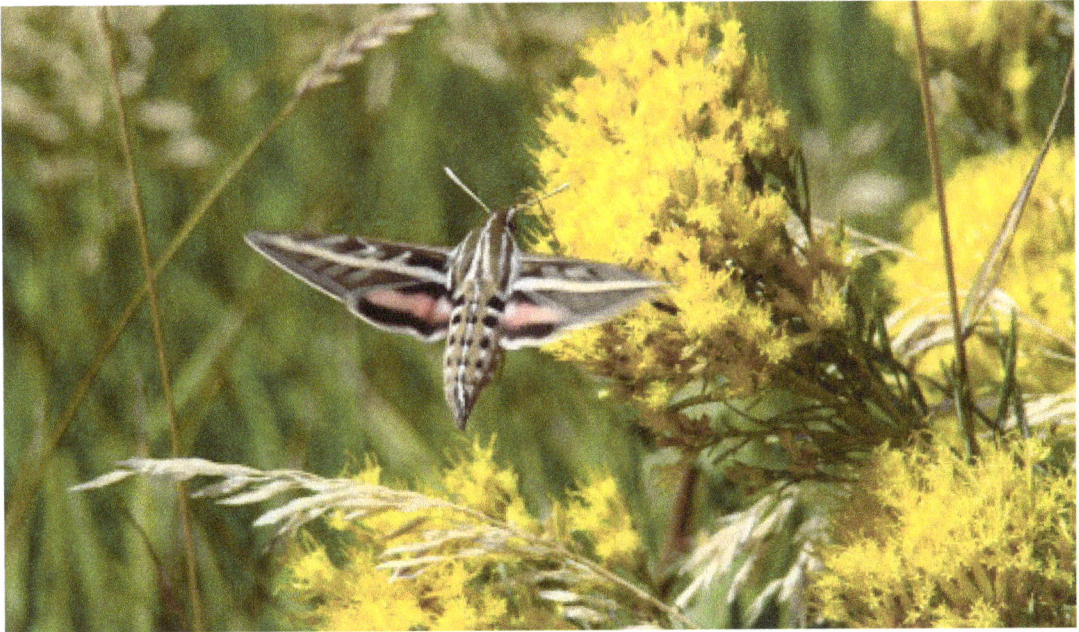

White-lined sphinx moth feeding in flight

Basic motion of the insect wing in insect with an indirect flight mechanism scheme of dorsoventral cut through a thorax segment with

a wings

b joints

c dorsoventral muscles

d longitudinal muscles.

Insects are the only group of invertebrates to have developed flight. The evolution of insect wings has been a subject of debate. Some entomologists suggest that the wings are from paranotal lobes, or extensions from the insect's exoskeleton called the nota, called the *paranotal theory*. Other theories are based on a pleural origin. These theories include suggestions that wings originated from modified gills, spiracular flaps or as from an appendage of the epicoxa. The *epicoxal theory* suggests the insect wings are modified epicoxal exites, a modified appendage at the base of the legs or coxa. In the Carboniferous age, some of the *Meganeura* dragonflies had as much as a 50 cm (20 in) wide wingspan. The appearance of gigantic insects has been found to be consistent with high atmospheric oxygen. The respiratory system of insects constrains their size, however the high oxygen in the atmosphere allowed larger sizes. The largest flying insects today are much smaller and include several moth species such as the Atlas moth and the White Witch (*Thysania agrippina*).

Insect flight has been a topic of great interest in aerodynamics due partly to the inability of steady-state theories to explain the lift generated by the tiny wings of insects. But insect wings are in motion, with flapping and vibrations, resulting in churning and eddies, and the misconception that physics says "bumblebees can't fly" persisted throughout most of the twentieth century.

Unlike birds, many small insects are swept along by the prevailing winds although many of the larger insects are known to make migrations. Aphids are known to be transported long distances by low-level jet streams. As such, fine line patterns associated with converging winds within weather radar imagery, like the WSR-88D radar network, often represent large groups of insects.

Walking

Many adult insects use six legs for walking and have adopted a tripedal gait. The tripedal gait allows for rapid walking while always having a stable stance and has been studied extensively in cockroaches. The legs are used in alternate triangles touching the ground. For the first step, the middle right leg and the front and rear left legs are in contact with the ground and move the insect forward, while the front and rear right leg and the middle left leg are lifted and moved forward to a new position. When they touch the ground to form a new stable triangle the other legs can be lifted and brought forward in turn and so on. The purest form of the tripedal gait is seen in insects moving at high speeds. However, this type of locomotion is not rigid and insects can adapt a variety of gaits. For example, when moving slowly, turning, or avoiding obstacles, four or more feet may be touching the ground. Insects can also adapt their gait to cope with the loss of one or more limbs.

Cockroaches are among the fastest insect runners and, at full speed, adopt a bipedal run to reach a high velocity in proportion to their body size. As cockroaches move very quickly, they need to be video recorded at several hundred frames per second to reveal their gait. More sedate locomotion is seen in the stick insects or walking sticks (Phasmatodea). A few insects have evolved to walk on the surface of the water, especially members of the Gerridae family, commonly known as water striders. A few species of ocean-skaters in the genus *Halobates* even live on the surface of open oceans, a habitat that has few insect species.

Use in Robotics

Insect walking is of particular interest as an alternative form of locomotion in robots. The study of insects and bipeds has a significant impact on possible robotic methods of transport. This may allow new robots to be designed that can traverse terrain that robots with wheels may be unable to handle.

Swimming

The backswimmer Notonecta glauca underwater, showing its paddle-like hindleg adaptation

A large number of insects live either part or the whole of their lives underwater. In many of the more primitive orders of insect, the immature stages are spent in an aquatic environment. Some groups of insects, like certain water beetles, have aquatic adults as well.

Many of these species have adaptations to help in under-water locomotion. Water beetles and water bugs have legs adapted into paddle-like structures. Dragonfly naiads use jet propulsion, forcibly expelling water out of their rectal chamber. Some species like the water striders are capable of walking on the surface of water. They can do this because their claws are not at the tips of the legs as in most insects, but recessed in a special groove further up the leg; this prevents the claws from piercing the water's surface film. Other insects such as the Rove beetle *Stenus* are known to emit pygidial gland secretions that reduce surface tension making it possible for them to move on the surface of water by Marangoni propulsion (also known by the German term *Entspannungsschwimmen*).

Ecology

Insect ecology is the scientific study of how insects, individually or as a community, interact with the surrounding environment or ecosystem.Insects play one of the most important roles in their ecosystems, which includes many roles, such as soil turning and aeration, dung burial, pest control, pollination and wildlife nutrition. An example is the beetles, which are scavengers that feed on dead animals and fallen trees and thereby recycle biological materials into forms found useful by other organisms. These insects, and others, are responsible for much of the process by which topsoil is created.

Defense and Predation

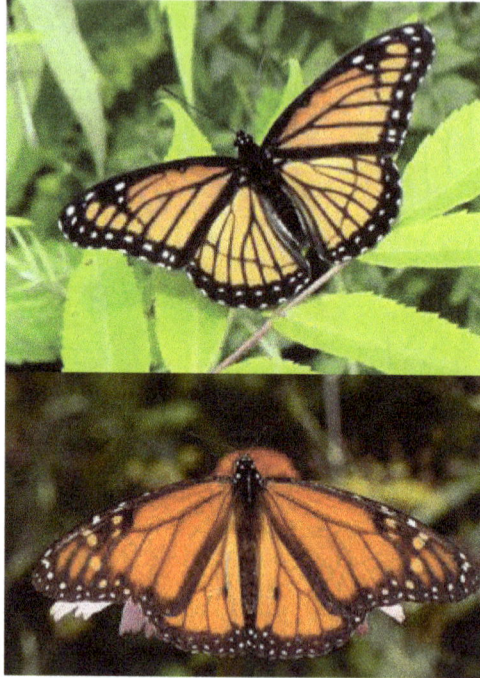

Perhaps one of the most well-known examples of mimicry, the viceroy butterfly (top) appears very similar to the noxious-tasting monarch butterfly (bottom).

Insects are mostly soft bodied, fragile and almost defenseless compared to other, larger lifeforms. The immature stages are small, move slowly or are immobile, and so all stages are exposed to predation and parasitism. Insects then have a variety of defense strategies to avoid being attacked by predators or parasitoids. These include camouflage, mimicry, toxicity and active defense.

Camouflage is an important defense strategy, which involves the use of coloration or shape to blend into the surrounding environment. This sort of protective coloration is common and widespread among beetle families, especially those that feed on wood or vegetation, such as many of the leaf beetles (family Chrysomelidae) or weevils. In some of these species, sculpturing or various colored scales or hairs cause the beetle to resemble bird dung or other inedible objects. Many of those that live in sandy environments blend in with the coloration of the substrate. Most phasmids are known for effectively replicating the forms of sticks and leaves, and the bodies of some species (such as *O. macklotti* and *Palophus centaurus*) are covered in mossy or lichenous outgrowths that supplement their disguise. Some species have the ability to change color as their surroundings shift (*B. scabrinota*, *T. californica*). In a further behavioral adaptation to supplement crypsis, a number of species have been noted to perform a rocking motion where the body is swayed from side to side that is thought to reflect the movement of leaves or twigs swaying in the breeze. Another method by which stick insects avoid predation and resemble twigs is by feigning death (catalepsy), where the insect enters a motionless state that can be maintained for a long period. The nocturnal feeding habits of adults also aids Phasmatodea in remaining concealed from predators.

Another defense that often uses color or shape to deceive potential enemies is mimicry. A number of longhorn beetles (family Cerambycidae) bear a striking resemblance to wasps, which helps them avoid predation even though the beetles are in fact harmless. Batesian and Müllerian mimicry complexes are commonly found in Lepidoptera. Genetic polymorphism and natural selection give rise to otherwise edible species (the mimic) gaining a survival advantage by resembling inedible species (the model). Such a mimicry complex is referred to as *Batesian* and is most commonly known by the mimicry by the limenitidine Viceroy butterfly of the inedible danaine Monarch. Later research has discovered that the Viceroy is, in fact more toxic than the Monarch and this resemblance should be considered as a case of Müllerian mimicry. In Müllerian mimicry, inedible species, usually within a taxonomic order, find it advantageous to resemble each other so as to reduce the sampling rate by predators who need to learn about the insects' inedibility. Taxa from the toxic genus *Heliconius* form one of the most well known Müllerian complexes.

Chemical defense is another important defense found amongst species of Coleoptera and Lepidoptera, usually being advertised by bright colors, such as the Monarch butterfly. They obtain their toxicity by sequestering the chemicals from the plants they eat into their own tissues. Some Lepidoptera manufacture their own toxins. Predators that eat poisonous butterflies and moths may become sick and vomit violently, learning not to eat those types of species; this is actually the basis of Müllerian mimicry. A predator who has previously eaten a poisonous lepidopteran may avoid other species with similar markings in the future, thus saving many other species as well. Some ground beetles of the Carabidae family can spray chemicals from their abdomen with great accuracy, to repel predators.

Pollination

European honey bee carrying pollen in a pollen basket back to the hive

Pollination is the process by which pollen is transferred in the reproduction of plants, thereby enabling fertilisation and sexual reproduction. Most flowering plants require an animal to do the transportation. While other animals are included as pollinators, the majority of pollination is done by insects. Because insects usually receive benefit for the pollination in the form of energy rich nectar it is a grand example of mutualism. The various flower traits (and combinations thereof) that differentially attract one type of pollinator or another are known as pollination syndromes. These arose through complex plant-animal adaptations. Pollinators find flowers through bright colorations, including ultraviolet, and attractant pheromones. The study of pollination by insects is known as *anthecology.*

Parasitism

Many insects are parasites of other insects such as the parasitoid wasps. These insects are known as entomophagous parasites. They can be beneficial due to their devastation of pests that can destroy crops and other resources. Many insects have a parasitic relationship with humans such as the mosquito. These insects are known to spread diseases such as malaria and yellow fever and because of such, mosquitoes indirectly cause more deaths of humans than any other animal.

Relationship to Humans

As Pests

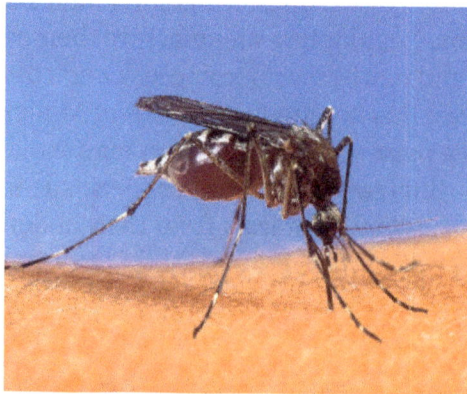

Aedes aegypti, a parasite, is the vector of dengue fever and yellow fever

Many insects are considered pests by humans. Insects commonly regarded as pests include those that are parasitic (*e.g.* lice, bed bugs), transmit diseases (mosquitoes, flies), damage structures (termites), or destroy agricultural goods (locusts, weevils). Many entomologists are involved in various forms of pest control, as in research for companies to produce insecticides, but increasingly rely on methods of biological pest control, or biocontrol. Biocontrol uses one organism to reduce the population density of another organism — the pest — and is considered a key element of integrated pest management.

Despite the large amount of effort focused at controlling insects, human attempts to kill pests with insecticides can backfire. If used carelessly, the poison can kill all kinds of organisms in the area, including insects' natural predators, such as birds, mice and other insectivores. The effects of DDT's use exemplifies how some insecticides can threaten wildlife beyond intended populations of pest insects.

In Beneficial Roles

Because they help flowering plants to cross-pollinate, some insects are critical to agriculture. This European honey bee is gathering nectar while pollen collects on its body.

Although pest insects attract the most attention, many insects are beneficial to the environment and to humans. Some insects, like wasps, bees, butterflies and ants, pollinate flowering plants. Pollination is a mutualistic relationship between plants and insects. As insects gather nectar from different plants of the same species, they also spread pollen from plants on which they have previously fed. This greatly increases plants' ability to cross-pollinate, which maintains and possibly even improves their evolutionary fitness. This ultimately affects humans since ensuring healthy crops is critical to agriculture. As well as pollination ants help with seed distribution of plants. This helps to spread the plants which increases plant diversity. This leads to an overall better environment. A serious environmental problem is the decline of populations of pollinator insects, and a number of species of insects are now cultured primarily for pollination management in order to have sufficient pollinators in the field, orchard or greenhouse at bloom time. Another solution, as shown in Delaware, has been to raise native plants to help support native pollinators like *L. vierecki*. Insects also produce useful substances such as honey, wax, lacquer and silk. Honey bees have been cultured by humans for thousands of years for honey, although contracting for crop pollination is becoming more significant for beekeepers. The silkworm has greatly affected human history, as silk-driven trade established relationships between China and the rest of the world.

Insectivorous insects, or insects which feed on other insects, are beneficial to humans because they eat insects that could cause damage to agriculture and human structures. For example, aphids feed on crops and cause problems for farmers, but ladybugs feed on aphids, and can be used as a means to get significantly reduce pest aphid populations. While birds are perhaps more visible predators of insects, insects themselves account for the vast majority of insect consumption. Ants also help control animal populations by consuming small vertebrates. Without predators to keep them in check, insects can undergo almost unstoppable population explosions.

A robberfly with its prey, a hoverfly. Insectivorous relationships such as these help control insect populations.

Insects are also used in medicine, for example fly larvae (maggots) were formerly used to treat wounds to prevent or stop gangrene, as they would only consume dead flesh. This treatment is finding modern usage in some hospitals. Recently insects have also gained attention as potential sources of drugs and other medicinal substances. Adult insects, such as crickets and insect larvae of various kinds, are also commonly used as fishing bait.

In Research

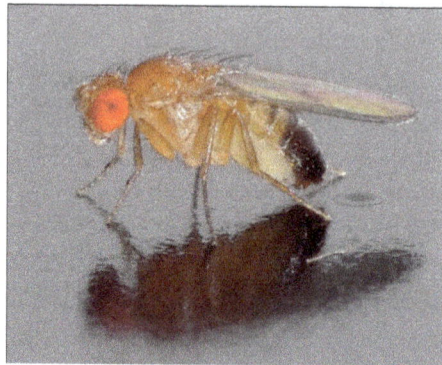

The common fruitfly Drosophila melanogaster is one of the most widely used organisms in biological research.

Insects play important roles in biological research. For example, because of its small size, short generation time and high fecundity, the common fruit fly *Drosophila melanogaster* is a model organism for studies in the genetics of higher eukaryotes. *D. melanogaster* has been an essential part of studies into principles like genetic linkage, interactions between genes, chromosomal genetics, development, behavior and evolution. Because genetic systems are well conserved among eukaryotes, understanding basic cellular processes like DNA replication or transcription in fruit flies can help to understand those processes in other eukaryotes, including humans. The genome of *D. mela-nogaster* was sequenced in 2000, reflecting the organism's important role in biological research. It was found that 70% of the fly genome is similar to the human genome, supporting the evolution theory.

As Food

In some cultures, insects, especially deep-fried cicadas, are considered to be delicacies, whereas in other places they form part of the normal diet. Insects have a high protein content for their mass, and some authors suggest their potential as a major source of protein in human nutrition. In most first-world countries, however, entomophagy (the eating of insects), is taboo. Since it is impossible to entirely eliminate pest insects from the human food chain, insects are inadvertently present in many foods, especially grains. Food safety laws in many countries do not prohibit insect parts in food, but rather limit their quantity. According to cultural materialist anthropologist Marvin Harris, the eating of insects is taboo in cultures that have other protein sources such as fish or livestock.

Due to the abundance of insects and a worldwide concern of food shortages, the Food and Agriculture Organisation of the United Nations considers that the world may have to, in the future, regard the prospects of eating insects as a food staple. Insects are not-ed for their nutrients, having a high content of protein, minerals and fats and are eaten by one-third of the global population.

In Culture

Scarab beetles held religious and cultural symbolism in Old Egypt, Greece and some shamanistic Old World cultures. The ancient Chinese regarded cicadas as symbols of rebirth or immortality. In Mesopotamian literature, the epic poem of Gilgamesh has allusions to Odonata which signify the impossibility of immortality. Amongst the Ab-origines of Australia of the Arrernte language groups, honey ants and witchety grubs served as personal clan totems. In the case of the 'San' bush-men of the Kalahari, it is the praying mantis which holds much cultural significance including creation and zen-like patience in waiting.

Beetle

Beetles are a group of insects that form the order Coleoptera. The front pair, the "ely-tra", being hardened and thickened into a shell-like protection for the rear pair and the beetle's abdomen. The order contains more species than any other order, constituting almost 25% of all known animal life-forms. About 40% of all described insect species are beetles (about 400,000 species), and new species are discovered frequently. The largest taxonomic family, the Curculionidae (the weevils or snout beetles), also belongs to this order.

The diversity of beetles is very wide-ranging. They are found in almost all types of habitats, but are not known to occur in the sea or in the polar regions. They interact with their ecosystems in several ways. They often feed on plants and fungi, break down animal and plant debris, and eat other invertebrates. Some species are prey of various animals including birds and mammals. Certain species are agricultural pests, such as the Colorado potato beetle *Leptinotarsa decemlineata*, the boll weevil *Anthonomus grandis*, the red flour beetle *Tribolium castaneum*, and the mungbean or cowpea beetle *Callosobruchus maculatus*, while other species of beetles are important controls of agricultural pests. For example, beetles in the family Coccinellidae ("ladybirds" or "ladybugs") consume aphids, scale insects, thrips, and other plant-sucking insects that damage crops.

Species in the order Coleoptera are generally characterized by a particularly hard exoskeleton and hard forewings (elytra, singular elytron). These elytra distinguish beetles from most other insect species, except for a few species of Hemiptera. The beetle's exoskeleton is made up of numerous plates called sclerites, separated by thin sutures. This design creates the armored defenses of the beetle while maintaining flexibility. The general anatomy of a beetle is quite uniform, although specific organs and appendages may vary greatly in appearance and function between the many families in the order. Like all insects, beetles' bodies are divided into three sections: the head, the thorax, and the abdomen. Coleopteran internal morphology is similar to other insects, although there are several examples of novelty. Such examples include species of water beetle which use air bubbles in order to dive under the water, and can remain submerged thanks to passive diffusion as oxygen moves from the water into the bubble.

Beetles are endopterygotes, which means that they undergo complete metamorphosis, a biological process by which an animal physically develops after birth or hatching, undergoing a series of conspicuous and relatively abrupt change in their body structure. Coleopteran species have an extremely intricate behavior when mating, using such methods as pheromones for communication to locate potential mates. Males may fight for females using very elongated mandibles, causing a strong divergence between males and females in sexual dimorphism.

Etymology

Coleoptera comes from the Greek *koleopteros*, literally "sheath-wing", from *koleos* meaning "sheath", and *pteron*, meaning "wing". The name was given to the group by Aristotle for their elytra, hardened shield-like forewings. The English name "beetle" comes from the Old English word *bitela*, literally meaning small biter, deriving from the word *bitel*, which means biting. This word is related to the word *bītan* (to bite) The name also derives from the Middle English word *betylle* from Old English *bitula* (also meaning to bite). Another Old English name for beetle is *ceafor*, chafer, used in names such as cockchafer, from the Proto-Germanic *kabraz- (compare German Käfer). These terms have been in use since the 12th century. In addition to names including the words "beetle" or "chafer", many groups of Coleoptera have common names such as fireflies, June bugs, ladybugs and weevils.

Taxonomy

A museum display, showing a little of the diversity of beetles

The Coleopterans include more species than any other order, constituting nearly 25% of all known types of animal life forms. About 450,000 species of beetles occur – representing about 40% of all known insects. Such a large number of species poses special problems for classification, with some families consisting of thousands of species and needing further division into subfamilies and tribes. This immense number of species allegedly led evolutionary biologist J. B. S. Haldane to quip, when some theologians asked him what could be inferred about the mind of the Creator from the works of His Creation, that God displayed "an inordinate fondness for beetles".

Polyphaga is the largest suborder, containing more than 300,000 described species in more than 170 families, including rove beetles (Staphylinidae), scarab beetles (Scarabaeidae), blister beetles (Meloidae), stag beetles (Lucanidae) and true weevils (Curculionidae). These beetles can be identified by the presence of cervical sclerites (hardened parts of the head used as points of attachment for muscles) absent in the other suborders. The suborder Adephaga contains about 10 families of largely predatory beetles, includes ground beetles (Carabidae), Dytiscidae and whirligig beetles

(Gyrinidae). In these beetles, the testes are tubular and the first abdominal sternum (a plate of the exoskeleton) is divided by the hind coxae (the basal joints of the beetle's legs). Archostemata contains four families of mainly wood-eating beetles, including reticulated beetles (Cupedidae) and the telephone-pole beetle. Myxophaga contains about 100 described species in four families, mostly very small, including Hydroscaphidae and the genus *Sphaerius*.

Evolution

The oldest known insect that unequivocally resembles species of Coleoptera date back to the Lower Permian (270 mya), though it instead has 13-segmented antennae, elytra with more fully developed venation and more irregular longitudinal ribbing, and an abdomen and ovipositor extending beyond the apex of the elytra. At the end of the Permian, the biggest mass extinction in history took place, collectively called the Permian–Triassic extinction event (P-Tr): 30% of all insect species became extinct; however, it is the only mass extinction of insects in Earth's history until today.

Due to the P-Tr extinction, the fossil record of insects only includes beetles from the Lower Triassic (220 million years ago). Around this time, during the Late Triassic, mycetophagous, or fungus-feeding species (e.g. Cupedidae) appear in the fossil record. In the stages of the Upper Triassic, representatives of the algophagous, or algae-feeding species (e.g. Triaplidae and Hydrophilidae) begin to appear, as well as predatory water beetles. The first primitive weevils appear (e.g. Obrienidae), as well as the first representatives of the rove beetles (e.g. Staphylinidae), which show no marked difference in morphology compared to recent species.

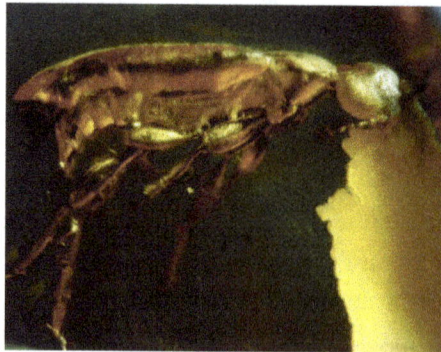

Baltic amber inclusions from the Eocene, 50 million years ago

During the Jurassic (210 to 145 million years ago), a dramatic increase in the known diversity of family-level Coleoptera occurred, including the development and growth of carnivorous and herbivorous species. Species of the superfamily Chrysomeloidea are believed to have developed around the same time, which include a wide array of plant hosts ranging from cycads and conifers, to angiosperms. Close to the Upper Jurassic, the portion of the Cupedidae decreased, but at the same time the diversity of the early plant-eating, or phytophagous species increased. Most of the recent phytophagous species of Coleoptera feed on flowering plants or angiosperms. The increase in diversity of the angiosperms is also believed to have influenced the diversity of the phytophagous species, which doubled during the Middle Jurassic. However, doubts have been raised recently, since the increase of the number of beetle families during the Cretaceous does not correlate with the increase of the number of angiosperm species. Also around the same time, numerous primitive weevils (e.g. Curculionoidea) and click beetles (e.g. Elateroidea) appeared. Also, the first jewel beetles (e.g. Buprestidae) are present, but they were rather rare until the Cretaceous. The first scarab beetles appeared around this time, but they were not coprophagous (feeding upon fecal matter), instead presumably feeding upon the rotting wood with the help of fungus; they are an early example of a mutualistic relationship.

The Cretaceous included the initiation of the most recent round of southern landmass fragmentation, via the opening of the southern Atlantic ocean and the isolation of New Zealand, while South America, Antarctica, and Australia grew more distant. During the Cretaceous, the diversity of Cupedidae and Archostemata decreased considerably. Predatory ground beetles (Carabidae) and rove beetles (Staphylinidae) began to distribute into different patterns; whereas the Carabidae predominantly occurred in the warm regions, the Staphylinidae and click beetles (Elateridae) preferred many areas with temperate climates. Likewise, predatory species of Cleroidea and Cucujoidea hunted their prey under the bark of trees together with the jewel beetles (Buprestidae). The jewel beetles' diversity increased rapidly during the Cretaceous, as they were the primary consumers of wood, while longhorn beetles (Cerambycidae) were rather rare, and their diversity increased only towards the end of the Upper Cretaceous. The first coprophagous beetles have been recorded from the Upper Cretaceous, and are believed to have lived on the excrement of herbivorous dinosaurs, but discussion is still ongoing as to whether the beetles were always tied to mammals during their development. Also, the first species with an adaption of both larvae and adults to the aquatic lifestyle are found. Whirligig beetles (Gyrinidae) were moderately diverse, although other early beetles (e.g. Dytiscidae) were less, with the most widespread being the species of Coptoclavidae, which preyed on aquatic fly larvae.

Between the Paleogene and the Neogene is when today's beetles developed. During this time, the continents began to be located closer to where they are today. Around 5 million years ago, the land bridge between South America and North America was formed, and the fauna exchange between Asia and North America started. Though many recent genera and species already existed during the Miocene, their distribution differed considerably from today's.

Fossil Record

A 2007 study based on DNA of living beetles and maps of likely beetle evolution indicated beetles may have originated during the Lower Permian, up to 285 million years ago. In 2009, a fossil beetle was described from the Pennsylvanian of Mazon Creek, Illinois, pushing the origin of the beetles to an earlier date, 318 to 299 million years ago.

Fossils from this time have been found in Asia and Europe, for instance in the red slate fossil beds of Niedermoschel near Mainz, Germany. Further fossils have been found in Obora, Czech Republic and Tshekarda in the Ural mountains, Russia. However, there are only a few fossils from North America before the middle Permian, although both Asia and North America had been united to Euramerica. The first discoveries from North America made in the Wellington formation of Oklahoma were published in 2005 and 2008.

Fossil buprestid beetle from the Eocene Messel pit, which retains its structural color

As a consequence of the Permian–Triassic extinction event, the fossil record of insects is scant, including beetles from the Lower Triassic. However, a few exceptions are noted, as in Eastern Europe; at the Babiy Kamen site in the Kuznetsk Basin, numerous beetle fossils were discovered, even entire specimen of the infraorders Archostemata (e.g. Ademosynidae, Schizocoleidae), Adephaga (e.., Triaplidae, Trachypachidae) and Polyphaga (e.g. Hydrophilidae, Byrrhidae, Elateroidea) and in nearly a perfectly preserved condition. However, species from the families Cupedidae and Schizophoroidae are not present at this site, whereas they dominate at other fossil sites from the Lower Triassic. Further records are known from Khey-Yaga, Russia, in the Korotaikha Basin. There are many important sites from the Jurassic, with more than 150 important sites with beetle fossils, the majority being situated in Eastern Europe and North Asia. In North America and especially in South America and Africa, the number of sites from that time period is smaller, and the sites have not been exhaustively investigated yet. Outstanding fossil sites include Solnhofen in Upper Bavaria, Germany, Karatau in South Kazakhstan, the Yixian formation in Liaoning, North China, as well as the Jiulongshan formation and further fossil sites in Mongolia. In North America there are only a few sites with fossil records of insects from the Jurassic, namely the shell limestone deposits in the Hartford basin, the Deerfield basin and the Newark basin.

A large number of important fossil sites worldwide contain beetles from the Cretaceous. Most are located in Europe and Asia and belong to the temperate climate zone during the Cretaceous. A few of the fossil sites mentioned in the chapter Jurassic also shed some light on the early Cretaceous beetle fauna (for example, the Yixian formation in Liaoning, North China). Further important sites from the Lower Cretaceous include the Crato fossil beds in the Araripe basin in the Ceará, North Brazil, as well as overlying Santana formation, with the latter was situated near the paleoequator, or the position of the earth's equator in the geologic past as defined for a specific geologic period. In Spain, important sites are located near Montsec and Las Hoyas. In Australia, the Koonwarra fossil beds of the Korumburra group, South Gippsland, Victoria, are noteworthy. Important fossil sites from the Upper Cretaceous include Kzyl-Dzhar in South Kazakhstan and Arkagala in Russia.

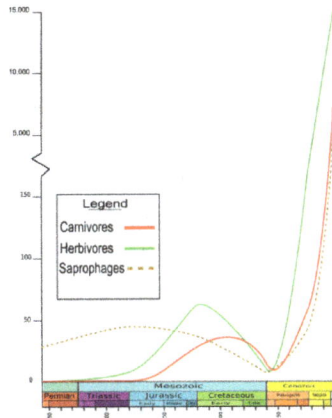

The phylogenetic growth of three different trophic levels in Coleoptera by number of genera

Phylogeny

The superficial consistency of most beetles' morphology, in particular their possession of elytra, has long suggested that Coleoptera is a monophyletic group. Growing evidence indicates this is unjustified, there being arguments for example, in favor of allocating the current suborder Adephaga their own order, or very likely even more than one. The suborders diverged in the Permian and Triassic. Their phylogenetic relationship is uncertain, with the most popular hypothesis being that Polyphaga and Myxophaga are most closely related, with Adephaga as the sister group to those two, and Archostemata as sister to the other three collectively. Although six other competing hypotheses are noted, the other most widely discussed one has Myxophaga as the sister group of all remaining beetles rather than just of Polyphaga. Evidence for a close relationship of the two suborders, Polyphaga and Myxophaga, includes the shared reduction in the number of larval leg articles. Adephaga is further considered as sister to Myxophaga and Polyphaga, based on their completely sclerotized elytra, reduced number of crossveins in the hind wings, and the folded (as opposed to rolled) hind wings of those three suborders.

Recent cladistic analysis of some of the structural characteristics supports the Polyphaga and Myxophaga hypothesis. The membership of the clade Coleoptera is not in dispute, with the exception of the twisted-wing parasites, Strepsiptera. These odd insects have been regarded as related to the beetle families Rhipiphoridae and Meloidae, with which they share first-instar larvae that are active, host-seeking triungulins and later-instar larvae that are endoparasites of other insects, or the sister group of beetles, or more distantly related to insects. Recent molecular genetic analysis strongly supports the hypothesis that Strepsiptera is the sister group to beetles.

Distribution and Diversity

Beetles are by far the largest order of insects, with 350,000–400,000 species in four sub-orders (Adephaga, Archostemata, Myxophaga, and Polyphaga), making up about 40% of all insect species described, and about 30% of all animals. Though classification at the family level is a bit unstable, about 500 families and subfamilies are recognized. One of the first proposed estimates of the total number of beetle species on the planet is based on field data rather than on catalog numbers. The technique used for this original estimate, possibly as many as 12 million species, was criticized, and was later revised, with estimates of 850,000–4,000,000 species proposed. Some 70–95% of all beetle species, depending on the estimate, remain undescribed. The beetle fauna is not equally well known in all parts of the world. For example, the known beetle diversity of Australia is estimated at 23,000 species in 3265 genera and 121 families. This is slightly lower than reported for North America, a land mass of similar size with 25,160 species in 3526 genera and 129 families. While other predictions show there could be as many as 28,000 species in North America, including those currently undescribed, a realistic estimate of the little-studied Australian beetle fauna's true diversity could vary from 80,000 to 100,000.

Coleoptera are found in nearly all natural habitats, including freshwater and marine habitats, everywhere vegetative foliage is found, from trees and their bark to flowers, leaves, and underground near roots- even inside plants in galls, in every plant tissue, including dead or decaying ones.

External Morphology

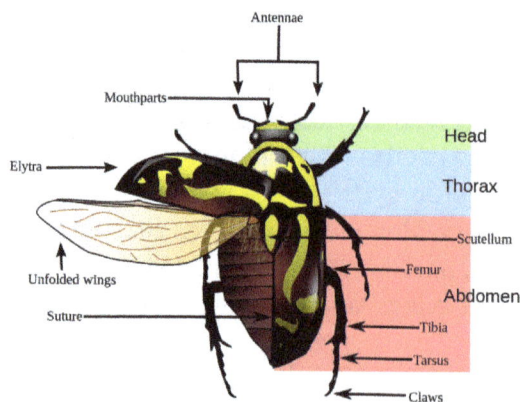

The morphology of a beetle, with a fiddler beetle as an example species

Beetles are generally characterized by a particularly hard exoskeleton and hard fore-wings (elytra). The beetle's exoskeleton is made up of numerous plates, called sclerites, separated by thin sutures. This design provides armored defenses while maintaining flexibility. The general anatomy of a beetle is quite uniform, although specific organs and appendages may vary greatly in appearance and function between the many families in the order. Like all insects, beetles' bodies are divided into three sections: the head, the thorax, and the abdomen.

Head

Scarabaeus viettei (syn. Madateuchus viettei, Scarabaeidae) showing a "shovel head" adaptation

The head, having mouthparts projecting forward or sometimes downturned, is usually heavily sclerotized and varies in size. The eyes are compound and may display remarkable adaptability, as in the case of whirligig beetles (family Gyrinidae), where they are split to allow a view both above and below the waterline. Other species also have divided eyes – some longhorn beetles (family Cerambycidae) and weevils – while many have eyes that are notched to some degree. A few beetle genera also possess ocelli, which are small, simple eyes usually situated farther back on the head (on the vertex).

Head of Cephalota circumdata, showing the compound eyes and mouthparts

Beetles' antennae are primarily organs of smell, but may also be used to feel a beetle's environment physically. They may also be used in some families during mating, or among a few beetle species for defence. Antennae vary greatly in form within the Coleoptera, but are often similar within any given family. Males and females sometimes have different antennal forms. Antennae may be clavate (flabellate and lamellate are subforms of clavate, or clubbed antennae), filiform, geniculate, moniliform, pectinate, or serrate.

Beetles have mouthparts similar to those of grasshoppers. Of these parts, the most commonly known are probably the mandibles, which appear as large pincers on the front of some beetles. The mandibles are a pair of hard, often tooth-like structures that move horizontally to grasp, crush, or cut food or enemies. Two pairs of finger-like appendages, the maxillary and labial palpi, are found around the mouth in most beetles, serving to move food into the mouth. In many species, the mandibles are sexually dimorphic, with the males' enlarged enormously compared with those of females of the same species.

Thorax

The thorax is segmented into the two discernible parts, the pro- and pterathorax. The pterathorax is the fused meso- and metathorax, which are commonly separated in other insect species, although flexibly articulate from the prothorax. When viewed from below, the thorax is that part from which all three pairs of legs and both pairs of wings arise. The abdomen is everything posterior to the thorax. When viewed from above, most beetles appear to have three clear sections, but this is deceptive: on the beetle's upper surface, the middle "section" is a hard plate called the pronotum, which is only the front part of the thorax; the back part of the thorax is concealed by the beetle's wings. This further segmentation is usually best seen on the abdomen.

Acilius sulcatus, a diving beetle showing hind legs adapted for life in water

Extremities

The multisegmented legs end in two to five small segments called tarsi. Like many other insect orders, beetles bear claws, usually one pair, on the end of the last tarsal segment of each leg. While most beetles use their legs for walking, legs may be variously modified and adapted for other uses. Among aquatic families – Dytiscidae, Haliplidae, many species of Hydrophilidae and others – the legs, most notably the last pair, are modified for swimming and often bear rows of long hairs to aid this purpose. Other beetles have fossorial legs that are widened and often spined for digging. Species with such adaptations are found among the scarabs, ground beetles, and clown beetles (family Histeridae). The hind legs of some beetles, such as flea beetles (within Chrysomelidae) and flea weevils (within Curculionidae), are enlarged and designed for jumping.

Wings

The elytra are connected to the pterathorax, so named because it is where the wings are connected (*pteron* meaning "wing" in Greek). The elytra are not used for flight, but tend to cover the hind part of the body and protect the second pair of wings (*alae*). They must be raised to move the hind flight wings. A beetle's flight wings are crossed with veins and are folded after landing, often along these veins, and stored below the elytra. A fold (*jugum*) of the membrane at the base of each wing is a characteristic feature. In some beetles, the ability to fly has been lost. These include some ground beetles (family Carabidae) and some "true weevils" (family Curculionidae), but also desert- and cave-dwelling species of other families. Many have the two elytra fused together, forming a solid shield over the abdomen. In a few families, both the ability to fly and the elytra have been lost, with the best known example being the glow-worms of the family Phengodidae, in which the females are larviform throughout their lives.

Abdomen

The abdomen is the section behind the metathorax, made up of a series of rings, each with a hole for breathing and respiration, called a spiracle, composing three different segmented sclerites: the tergum, pleura, and the sternum. The tergum in almost all species is membranous, or usually soft and concealed by the wings and elytra when not in flight. The pleura are usually small or hidden in some species, with each pleuron having a single spiracle. The sternum is the most widely visible part of the abdomen, being a more or less sclerotized segment. The abdomen itself does not have any appendages, but some (for example, Mordellidae) have articulating sternal lobes.

Internal Morphology

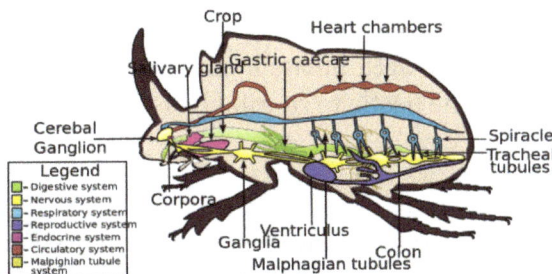

A diagram showing the general internal anatomy of beetles

Digestive System

The digestive system of beetles is primarily based on plants, upon which they, for the most part, feed, with mostly the anterior midgut performing digestion, although in predatory species (for example Carabidae), most digestion occurs in the crop by means of midgut enzymes. In Elateridae species, the predatory larvae defecate enzymes on their prey, with digestion being extraorally. The alimentary canal basically consists of a short, narrow pharynx, a widened expansion, the crop, and a poorly developed gizzard. After is the midgut, that varies in dimensions between species, with a large amount of cecum, with a hindgut, with varying lengths. Typically, four to six Malpighian tubules occur.

Nervous System

The nervous system in beetles contains all the types found in insects, varying between different species, from three thoracic and seven or eight abdominal ganglia which can be distinguished to that in which all the thoracic and abdominal ganglia are fused to form a composite structure.

Respiratory System

Like most insects, beetles inhale oxygen and exhale carbon dioxide via a tracheal system. Air enters the body through spiracles, and circulates within the haemocoel in a system of tracheae and tracheoles, through the walls of which the relevant gases can diffuse appropriately.

Diving beetles, such as the Dytiscidae, carry a bubble of air with them when they dive. Such a bubble may be contained under the elytra or against the body by specialized hydrophobic hairs. The bubble covers at least some of the spiracles, thereby permitting the oxygen to enter the tracheae.

The function of the bubble is not so much as to contain a store of air, but to act as a physical gill. The air that it traps is in contact with oxygenated water, so as the animal's consumption depletes the oxygen in the bubble, more oxygen can diffuse in to replenish it. Carbon dioxide is more soluble in water than either oxygen or nitrogen, so it readily diffuses out faster than in. Nitrogen is the most plentiful gas in the bubble, and the least soluble, so it constitutes a relatively static component of the bubble and acts as a stable medium for respiratory gases to accumulate in and pass through. Occasional visits to the surface are sufficient for the beetle to re-establish the constitution of the bubble.

Circulatory System

Like other insects, beetles have open circulatory systems, based on hemolymph rather than blood. Also as in other insects, a segmented tube-like heart is attached to the dorsal wall of the hemocoel. It has paired inlets or *ostia* at intervals down its length, and circulates the hemolymph from the main cavity of the haemocoel and out through the anterior cavity in the head.

Specialized Organs

Different glands specialize for different pheromones produced for finding mates. Pheromones from species of Rutelinea are produced from epithelial cells lining the inner surface of the apical abdominal segments; amino acid-based pheromones of Melolonthinae are produced from eversible glands on the abdominal apex. Other species produce different types of pheromones. Dermestids produce esters, and species of Elateridae produce fatty acid-derived aldehydes and acetates. For means of finding a mate also, fireflies (Lampyridae) use modified fat body cells with transparent surfaces backed with reflective uric acid crystals to biosynthetically produce light, or bioluminescence. The light produce is highly efficient, as it is produced by oxidation of luciferin by enzymes (luciferases) in the presence of adenosine triphosphate (ATP) and oxygen, producing oxyluciferin, carbon dioxide, and light.

A notable number of species have developed special glands to produce chemicals for deterring predators. The ground beetle's (of Carabidae) defensive glands, located at the posterior, produce a variety of hydrocarbons, aldehydes, phenols, quinones, esters, and acids released from an opening at the end of the abdomen. African carabid beetles (for example, *Anthia* and *Thermophilum* – Thermophilum generally included within *Anthia*) employ the same chemicals as ants: formic acid. Bombardier beetles have well-developed, like other carabid beetles, pygidial glands that empty from the lateral edges of the intersegment membranes between the seventh and eighth abdominal segments. The gland is made of two containing chambers. The first holds hydroquinones and hydrogen peroxide, with the second holding just hydrogen peroxide plus catalases. These chemicals mix and result in an explosive ejection, forming temperatures of around 100 °C (212 °F), with the breakdown of hydroquinone to $H_2 + O_2$ + quinone, with the O_2 propelling the excretion.

Tympanal organs or hearing organs, which is a membrane (tympanum) stretched across a frame backed by an air sac and associated sensory neurons, are described in two families. Several species of the genus *Cicindela* (Cicindelidae) have ears on the dorsal surfaces of their first abdominal segments beneath the wings; two tribes in the subfamily Dynastinae (Scarabaeidae) have ears just beneath their pronotal shields or neck membranes. The ears of both families are sensitive to ultrasonic frequencies, with strong evidence indicating they function to detect the presence of bats by their ultrasonic echolocation. Though beetles constitute a large order and live in a variety of niches, examples of hearing are surprisingly lacking amongst species, though likely most simply remain undiscovered.

Reproduction and Development

Beetles are members of the superorder Endopterygota, and accordingly most of them undergo complete metamorphosis. The typical form of metamorphosis in beetles passes through four main stages: the egg, the larva, the pupa, and the imago or adult. The larvae are commonly called grubs and the pupa sometimes is called the chrysalis. In some species, the pupa may be enclosed in a cocoon constructed by the larva towards the end of its final instar. Going beyond "complete metamorphosis", however, some beetles, such as typical members of the families Meloidae and Rhipiphoridae, undergo hypermetamorphosis in which the first instar takes the form of a triungulin.

Mating

Punctate flower chafers (*Neorrhina punctata*, Scarabaeidae) mating

Beetles may display extremely intricate behavior when mating. Pheromone communication is likely to be important in the location of a mate.

Different species use different chemicals for their pheromones. Some scarab beetles (for example, Rutelinae) utilize pheromones derived from fatty acid synthesis, while other scarab beetles use amino acids and terpenoid compounds (for example, Melolonthinae). Another way species of Coleoptera find mates is the use of biosynthesized light, or bioluminescence. This special form of a mating call is confined to fireflies (Lampyridae) by the use of abdominal light-producing organs. The males and females engage in complex dialogue before mating, identifying different species by differences in duration, flight patterns, composition, and intensity.

Before mating, males and females may engage in various forms of behavior. They may stridulate, or vibrate the objects they are on. In some species (for example, Meloidae), the male climbs onto the dorsum of the female and strokes his antennae on her head, palps, and antennae. In the genus *Eupompha* of said family, the male draws the antennae along his longitudinal vertex. They may not mate at all if they do not perform the precopulatory ritual.

Conflict can play a part in the mating rituals of species such as burying beetles (genus *Nicrophorus*), where conflicts between males and females rage until only one of each is left, thus ensuring reproduction by the strongest and fittest. Many male beetles are territorial and fiercely defend their small patches of territory from intruding males. In such species, the male often has horns on the head or thorax, making its body length greater than that of a female. Pairing is generally quick, but in some cases lasts for several hours. During pairing, sperm cells are transferred to the female to fertilize the egg.

Lifecycle

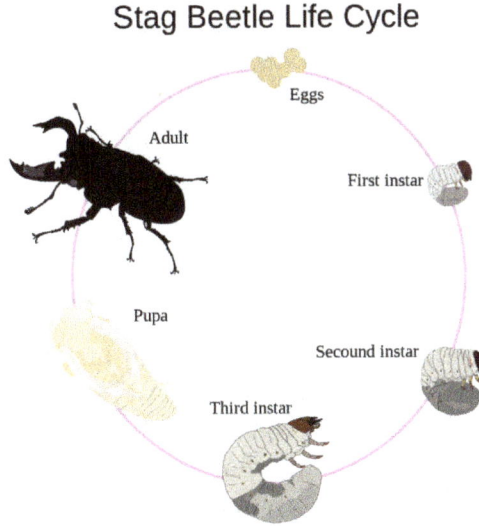

The lifecycle of the stag beetle includes three instars.

Egg

A single female may lay from several dozen to several thousand eggs during her lifetime. Eggs are usually laid according to the substrate on which the larvae feed upon hatching. Among others, they can be laid loose in the substrate (for example, flour beetle), laid in clumps on leaves (for example, Colorado potato beetle), individually attached (for example, mungbean beetle and other seed borers), or buried in the medium (for example, carrot weevil).

Parental care varies between species, ranging from the simple laying of eggs under a leaf to certain scarab beetles, which construct underground structures complete with a supply of dung to house and feed their young. Other beetles are leaf rollers, biting sections of leaves to cause them to curl inwards, then laying their eggs, thus protected, inside.

Larva

The larva is usually the principal feeding stage of the beetle lifecycle. Larvae tend to feed voraciously once they emerge from their eggs. Some feed externally on plants, such as those of certain leaf beetles, while others feed within their food sources. Examples of internal feeders are most Buprestidae and longhorn beetles. The larvae of many beetle families are predatory like the adults (ground beetles, ladybirds, rove beetles). The larval period varies between species, but can be as long as several years. The larvae are highly varied amongst species, with well-developed and sclerotized heads, and have distinguishable thoracic and abdominal segments (usually the tenth, though sometimes the eighth or ninth).

A scarabaeiform larva known as a curl grub

Beetle larvae can be differentiated from other insect larvae by their hardened, often darkened heads, the presence of chewing mouthparts, and spiracles along the sides of their bodies. Like adult beetles, the larvae are varied in appearance, particularly between beetle families. Beetles with somewhat flattened, highly mobile larvae include the ground beetles, some rove beetles, and others; their larvae are described as campodeiform. Some beetle larvae resemble hardened worms with dark head capsules and minute legs. These are elateriform larvae, and are found in the click beetle (Elateridae) and darkling beetle (Tenebrionidae) families. Some elateriform larvae of click beetles are known as wireworms. Beetles in the Scarabaeoidea have short, thick larvae described as scarabaeiform, more commonly known as grubs.

All beetle larvae go through several instars, which are the developmental stages between each moult. In many species, the larvae simply increase in size with each successive instar as more food is consumed. In some cases, however, more dramatic changes occur. Among certain beetle families or genera, particularly those that exhibit parasitic lifestyles, the first instar (the planidium) is highly mobile to search out a host, while the following instars are more sedentary and remain on or within their host. This is known as hypermetamorphosis; examples include the blister beetles (family Meloidae) and some rove beetles, particularly those of the genus *Aleochara*.

Pupa

As with all endopterygotes, beetle larvae pupate, and from these pupae emerge fully formed, sexually mature adult beetles, or imagos. Adults have extremely variable lifespans, from weeks to years, depending on the species. In some species, the pupa may go through all four forms during its development, called hypermetamorphosis (for example, Meloidae). Pupae always have no mandibles (are adecticous). In most, the appendages are not attached to the pupae; ones that do have appendages are mostly obtect, and the rest are exarate.

Behavior

Locomotion

Photinus pyralis, firefly, in flight

The elytra allow beetles to both fly and move through confined spaces, doing so by folding the delicate wings under the elytra while not flying, and folding their wings out just before take off. The unfolding and folding of the wings is operated by muscles attached to the wing base; as long as the tension on the radial and cubital veins remains, the wings remain straight. In day-flying species (for example, Buprestidae, Scarabaeidae), flight does not include large amounts of lifting of the elytra, having the metathorac wings extended under the lateral elytra margins.

Aquatic beetles use several techniques for retaining air beneath the water's surface. Beetles of the family Dytiscidae hold air between the abdomen and the elytra when diving. Hydrophilidae have hairs on their under surface that retain a layer of air against their bodies. Adult crawling water beetles use both their elytra and their hind coxae (the basal segment of the back legs) in air retention, while whirligig beetles simply carry an air bubble down with them whenever they dive.

Communication

Beetles have a variety of ways to communicate, some of which include a sophisticated chemical language through the use of pheromones. The pheromone language of the mountain pine beetle is well known to scientists. In order to overcome the chemical defenses of potential host trees, the beetles emit a pheromone that attracts other beetles to the tree. This 'call for help' allows the beetles to colonize the tree. After the tree's defenses have been exhausted and many beetles have arrived at the tree, the mountain pine beetle emits an anti-aggregation pheromone which makes the tree no longer attractive. This helps to avoid the harmful effects of having too many beetles on one tree competing for resources. The mountain pine beetle can also stridulate to communicate, or rub body parts together to create sound, having a "scraper" on their abdomens that they rub against a grooved surface on the underside of their left wing cover to create a sound that is not audible to humans. Once the female beetles have arrived on a suitable pine tree host, they begin to stridulate and produce aggregation pheromones to attract other unmated males and females. New females arrive and do the same as they land and bore into the tree. As the males arrive, they enter the galleries that the females have tunneled, and begin to stridulate to let the females know they have arrived, and to also warn others that the female in that gallery is taken. At this point, the female stops producing aggregation pheromones and starts producing anti-aggregation pheromone to deter more beetles from coming.

The environment and local climate can affect communication by beetles. Temperature, humidity, and wind speed can influence how pheromones travel through the air. For instance, high wind speeds may make pheromones difficult to detect by beetles locating for mates. Stridulation can be interrupted by noise in the environment.

Parental Care

A dung beetle rolling dung

Among insects, parental care is very uncommon, only found in a few species. Some beetles also display this unique social behavior. One theory states parental care is necessary for the survival of the larvae, protecting them from adverse environmental conditions and predators. One species, a rover beetle (*Bledius spectabilis*) displays both causes for parental care: physical and biotic environmental factors. Said species lives in salt marshes, so the eggs and/or larvae are endangered by the rising tide. The maternal beetle patrols the eggs and larvae and applies the appropriate burrowing behavior to keep them from flooding and from asphyxiating. Another advantage is that the mother protects the eggs and larvae from the predatory carabid beetle *Dicheirotrichus gustavi* and from the parasitoid wasp *Barycnemis blediator*. Up to 15% of larvae are killed by this parasitoid wasp, being only protected by maternal beetles in their dens.

Some species of dung beetle also display a form of parental care. Dung beetles collect animal feces, or "dung", from which their name is derived, and roll it into a ball, sometimes being up to 50 times their own weight; albeit sometimes it is also used to store food. Usually it is the male that rolls the ball, with the female hitch-hiking or simply following behind. In some cases the male and the female roll together. When a spot with soft soil is found, they stop and bury the dung ball. They then mate underground. After the mating, one or both of them prepares the brooding ball. When the ball is finished, the female lays eggs inside it, a form of mass provisioning. Some species do not leave after this stage, but remain to safeguard their offspring.

Mylabris pustulata (Meloidae) feeding on the petals of Ipomoea carnea

Feeding

Besides being abundant and varied, beetles are able to exploit the wide diversity of food sources available in their many habitats. Some are omnivores, eating both plants and animals. Other beetles are highly specialized in their diet. Many species of leaf beetles, longhorn beetles, and weevils are very host-specific, feeding on only a single species of plant. Ground beetles and rove beetles (family Staphylinidae), among others, are primarily carnivorous and catch and consume many other arthropods and small prey, such as earthworms and snails. While most predatory beetles are generalists, a few species have more specific prey requirements or preferences.

Decaying organic matter is a primary diet for many species. This can range from dung, which is consumed by coprophagous species (such as certain scarab beetles of the family Scarabaeidae), to dead animals, which are eaten by necrophagous species (such as the carrion beetles of the family Silphidae). Some of the beetles found within dung and carrion are in fact predatory. These include the clown beetles, preying on the larvae of coprophagous and necrophagous insects.

Ecology

Defense and Predation

Beetles and their larvae have a variety of strategies to avoid being attacked by predators or parasitoids. These include camouflage, mimicry, toxicity, and active defense. Camouflage involves the use of coloration or shape to blend into the surrounding environment. This sort of protective coloration is common and widespread among beetle families, especially those that feed on wood or vegetation, such as many of the leaf beetles (family Chrysomelidae) or weevils. In some of these species, sculpturing or various colored scales or hairs cause the beetle to resemble bird dung or other inedible objects. Many of those that live in sandy environments blend in with the coloration of the substrate. The giant African longhorn beetle (*Petrognatha gigas*) resembles the moss and bark of the tree it feeds on. Another defense that often uses color or shape to deceive potential enemies is mimicry. Some longhorn beetles (family Cerambycidae) bear a striking resemblance to wasps, which helps them avoid predation even though the beetles are in fact harmless. This defense is an example of Batesian mimicry and, together with other forms of mimicry and camouflage occurs widely in other beetle families, such as the Scarabaeidae. Beetles may combine their color mimicry with behavioral mimicry, acting like the wasps they already closely resemble. Many beetle species, including ladybirds, blister beetles, and lycid beetles can secrete distasteful or toxic substances to make them unpalatable or even poisonous. These same species are often aposematic, where bright or contrasting color patterns warn away potential predators; many beetles and other insects mimic these chemically protected species.

Beetles may be preyed upon by other insects such as robber flies

Chemical defense is another important defense found amongst species of Coleoptera, usually being advertised by bright colors. Others may utilize behaviors that would be done when releasing noxious chemicals (for example, Tenebrionidae). Chemical defense may serve purposes other than just protection from vertebrates, such as protection from a wide range of microbes, and repellents. Some species release chemicals in the form of a spray with surprising accuracy, such as ground beetles (Carabidae), may spray chemicals from their abdomen to repel predators. Some species take advantage of the plants from which they feed, and sequester the chemicals from the plant that would protect it and incorporate into their own defense. African carabid beetles (for example, *Anthia* and *Thermophilum*) employ the same chemicals used by ants, while bombardier beetles have a their own unique separate gland, spraying potential predators from far distances.

Clytus arietis (Cerambycidae), a wasp mimic

Large ground beetles and longhorn beetles may defend themselves using strong mandibles, spines or horns to forcibly persuade a predator to seek out easier prey. Many species such as the rhinoceros beetle have large protrusions from their thorax and head, which can be used to defend themselves from predators. Many species of weevil that feed out in the open on leaves of plants react to attack by employing a "drop-off reflex". Some combine it with thanatosis, in which they close up their appendages and "play dead".

Parasitism

Over 1000 species of beetles are known to be either parasitic, predatory, or commensals in the nests of ants.

A few species of beetles are actually ectoparasitic on mammals. One such species, *Platypsyllus castoris*, parasitises beavers (*Castor* spp.). This beetle lives as a parasite both as a larva and as an adult, feeding on epidermal tissue and possibly on skin secretions and wound exudates. They are strikingly flattened dorsoventrally, no doubt as an adaptation for slipping between the beavers' hairs. They also are wingless and eyeless, as are many other ectoparasites.

Other parasitic beetles include those that are kleptoparasites of other invertebrates, such as the small hive beetle (*Aethina tumida*) that infests honey bee hives. The larvae tunnel through comb towards stored honey or pollen, damaging or destroying cappings and comb in the process. Larvae defecate in honey and the honey becomes discolored from the feces, which causes fermentation and a frothiness in the honey; the honey develops a characteristic odor of decaying oranges. Damage and fermentation cause honey to run out of combs, destroying large amounts of it both in hives and sometimes also in honey extracting rooms. Heavy infestations cause bees to abscond; though the beetle is only a minor pest in Africa, beekeepers in other regions have reported the rapid collapse of even strong colonies.

Pollination

Beetle-pollinated flowers are usually large, greenish or off-white in color, and heavily scented. Scents may be spicy, fruity, or similar to decaying organic material. Most beetle-pollinated flowers are flattened or dish-shaped, with pollen easily accessible, although they may include traps to keep the beetle longer. The plants' ovaries are usually well protected from the biting mouthparts of their pollinators. Beetles may be particularly important in some parts of the world such as semiarid areas of southern Africa and southern California and the montane grasslands of KwaZulu-Natal in South Africa.

Beetle pollinates the flower

Mutualism

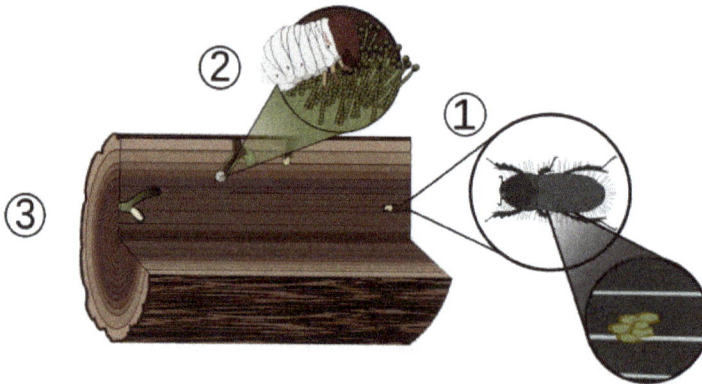

1: The adult beetle burrows hole into wood and lays eggs, carrying fungal spores in its mycangia.
2: The larva feeds on the fungus, which digest the wood, removing toxins: they mutu-ally benefit.
3: The larva pupates and then ecloses.

Amongst most orders of insects, mutualism is not common, but some examples occur in species of Coleoptera, such as the ambrosia beetle, the ambrosia fungus, and probably bacteria. The beetles excavate tunnels in dead trees in which they cultivate fungal gardens, their sole source of nutrition. After landing on a suitable tree, an ambrosia beetle excavates a tunnel in which it releases spores of its fungal symbiont. The fungus penetrates the plant's xylem tissue, digests it, and concentrates the nutrients on and near the surface of the beetle gallery, so the weevils and the fungus both benefit. The beetles cannot eat the wood due to toxins, and uses its relationship with fungi to help overcome its host tree defenses and to provide nutrition for their larvae. Chemically mediated by a bacterially produced polyunsaturated peroxide, this mutualistic relationship between the beetle and the fungus is coevolved.

Commensalism

Pseudoscorpions are small arachnids with flat, pear-shaped bodies and pincers that resemble those of scorpions (only distant relatives), usually ranging from 2 to 8 millimetres (0.08 to 0.31 in) in length. Their small size allows them to hitch rides under the elytra of giant harlequin beetles to be dispersed over wide areas while simultaneously being protected from predators. They may also find mating partners as other individuals join them on the beetle. This would be a form of parasitism if the beetle were harmed in the process, but the beetle is, presumably, unaffected by the presence of the hitchhikers.

Eusociality

Austroplatypus incompertus is eusocial, one of the few organisms outside Hymenoptera to do so, and the only species of Coleoptera.

Relationship to Humans

As Pests

Cotton boll weevil

About 75% of beetle species are phytophagous in both the larval and adult stages, and live in or on plants, wood, fungi, and a variety of stored products, including cereals, tobacco, and dried fruits. Because many of these plants are important for agriculture, forestry, and the household, beetles can be considered pests. Some of these species cause significant damage, such as the boll weevil, which feeds on cotton buds and flowers. The boll weevil crossed the Rio Grande near Brownsville, Texas, to enter the United States from Mexico around 1892, and had reached southeastern Alabama by 1915. By the mid-1920s, it had entered all cotton-growing regions in the US, traveling 40 to 160 miles (60–260 km) per year. It remains the most destructive cotton pest in North America. Mississippi State University has estimated, since the boll weevil entered the United States, it has cost cotton producers about $13 billion, and in recent times about $300 million per year. Many other species also have done extensive damage to plant populations, such as the bark beetle, elm leaf beetle and Asian longhorned beetle. The bark beetle, elm leaf beetle and Asian longhorned beetle, among other species, have been known to nest in elm trees. Bark beetles in particular carry Dutch elm disease as they move from infected breeding sites to feed on healthy elm trees, which in turn allows the Asian longhorned beetle to continue killing more elms. The spread of Dutch elm disease by the beetle has led to the devastation of elm trees in many parts of the Northern Hemisphere, notably in Europe and North America.

Larvae of the Colorado potato beetle, Leptinotarsa decemlineata

Situations in which a species has developed immunity to pesticides are worse, as in the case of the Colorado potato beetle, *Leptinotarsa decemlineata*, which is a notorious pest of potato plants. Crops are destroyed and the beetle can only be treated by employing expensive pesticides, to many of which it has begun to develop resistance. Suitable hosts can include a number of plants from the potato family (Solanaceae), such as nightshade, tomato, eggplant and capsicum, as well as potatoes. The Colorado potato beetle has developed resistance to all major insecticide classes, although not every population is resistant to every chemical.

Pests do not only affect agriculture, but can also even affect houses, such as the death watch beetle. The death watch beetle, *Xestobium rufovillosum* (family Anobiidae), is of considerable importance as a pest of older wooden buildings in Great Britain. It attacks hardwoods such as oak and chestnut, always where some fungal decay has taken or is taking place. The actual introduction of the pest into buildings is thought to take place at the time of construction.

Other pest include the coconut hispine beetle, *Brontispa longissima*, which feeds on young leaves and damages seedlings and mature coconut palms. On September 27, 2007, Philippines' Metro Manila and 26 provinces were quarantined due to having been infested with this pest (to save the $800-million Philippine coconut industry). The mountain pine beetle normally attacks mature or weakened lodgepole pine. It can be the most destructive insect pest of mature pine forests. The current infestation in British Columbia is the largest Canada has ever seen.

As Beneficial Resources

Coccinella septempunctata, a beneficial beetle

Beetles are not only pests, but can also be beneficial, usually by controlling the populations of pests. One of the best, and widely known, examples are the ladybugs or ladybirds (family Coccinellidae). Both the larvae and adults are found feeding on aphid colonies. Other ladybugs feed on scale insects and mealybugs. If normal food sources are scarce, they may feed on small caterpillars, young plant bugs, or honeydew and nectar. Ground beetles (family Carabidae) are common predators of many different insects and other arthropods, including fly eggs, caterpillars, wireworms, and others.

Dung beetles (Scarabidae) have been successfully used to reduce the populations of pestilent flies and parasitic worms that breed in cattle dung. The beetles make the dung unavailable to breeding pests by quickly rolling and burying it in the soil, with the added effect of improving soil fertility, tilth, and nutrient cycling. The Australian Dung Beetle Project (1965–1985), led by Dr. George Bornemissza of the Commonwealth Scientific and Industrial Research Organisation, introduced species of dung beetle to Australia from South Africa and Europe, and effectively reduced the bush fly (*Musca vetustissima*) population by 90%.

Dung beetles play a remarkable role in agriculture. By burying and consuming dung, they improve nutrient recycling and soil structure. They also protect livestock, such as cattle, by removing dung, which, if left, could provide habitat for pests such as flies. Therefore, many countries have introduced the creatures for the benefit of animal husbandry. In developing countries, the beetle is especially important as an adjunct for improving standards of hygiene. The American Institute of Biological Sciences reports that dung beetles save the United States cattle industry an estimated US$380 million annually through burying above-ground livestock feces.

Some beetles help in a professional setting, doing things that people cannot; those of the family Dermestidae are often used in taxidermy and preparation of scientific specimens to clean bones of remaining soft tissue. The beetle larvae are used to clean skulls because they do a thorough job of cleaning, and do not leave the tool marks that taxidermists' tools do. Another benefit is, with no traces of meat remaining and no emulsified fats in the bones, the trophy does not develop the unpleasant dead odor. Using the beetle larvae means that all cartilage is removed along with the flesh, leaving the bones spotless.

Mealworms presented in a bowl for human consumption

As Food

Insects are used as human food in 80% of the world's nations. Beetles are the most widely eaten insects. About 344 species are known to be used as food, usually eaten in the larval stage. The mealworm is the most commonly eaten beetle species. The larvae of the darkling beetle and the rhinoceros beetle are also commonly eaten.

In Art

Zopheridae examples of jewelry taken at the Texas A&M University Insect Collection in College Station, Texas

Many beetles have beautiful and durable elytra that have been used as material in arts, with beetlewing the best example. Sometimes, they are also incorporated into ritual objects for their religious significance. Whole beetles, either as-is or encased in clear plastic, are also made into objects varying from cheap souvenirs such as key chains to expensive fine-art jewelry. In parts of Mexico, beetles of the genus *Zopherus* are made into living brooches by attaching costume jewelry and golden chains, which is made possible by the incredibly hard elytra and sedentary habits of the genus.

In Ancient Cultures

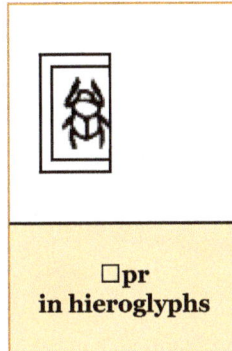

□pr
in hieroglyphs

Some beetles were prominent in ancient cultures, the most prominent being the dung beetle in Ancient Egypt. Several species of dung beetle, especially the "sacred scarab" *Scarabaeus sacer*, were revered by the ancient Egyptians. The hieroglyphic image of the beetle may have had existential, fictional, or ontologic significance. Images of the scarab in bone, ivory, stone, Egyptian faience, and precious metals are known from the Sixth Dynasty and up to the period of Roman rule. The scarab was of prime significance in the funerary cult of ancient Egypt.

The scarab was linked to Khepri, the god of the rising sun, from the supposed resemblance of the dung ball rolled by the beetle to the rolling of the sun by the god. Plutarch wrote:

The race of beetles has no female, but all the males eject their sperm into a round pellet of material which they roll up by pushing it from the opposite side, just as the sun seems to turn the heavens in the direction opposite to its own course, which is from west to east.

In contrast to funerary contexts, some of ancient Egypt's neighbors adopted the scarab motif for seals of varying types. The best-known of these are the Judean LMLK seals (eight of 21 designs contained scarab beetles), which were used exclusively to stamp impressions on storage jars during the reign of Hezekiah.

A scarab statue in the Karnak temple complex

A scarab on a wall of Tomb KV6 in the Valley of the Kings

In Modern Cultures

Beetles still play roles in culture. One example is in insect fighting for entertainment and gambling. This sport exploits the territorial behavior and mating competition of certain species of large beetles. In the Chiang Mai district of northern Thailand, male *Xylotrupes* rhinoceros beetles are caught in the wild and trained for fighting. Females are held inside a log to stimulate the fighting males with their pheromones.

Hemiptera

The Hemiptera or true bugs are an order of insects comprising around 50,000–80,000 species of groups such as the cicadas, aphids, planthoppers, leafhoppers, and shield bugs. They range in size from 1 mm (0.04 in) to around 15 cm (6 in), and share a common arrangement of sucking mouthparts. The name "true bugs" is sometimes limited to the suborder Heteroptera. Many insects commonly known as "bugs" belong to other orders; for example, the lovebug is a fly, while the May bug and ladybug are beetles.

Most hemipterans feed on plants, using their sucking and piercing mouthparts to extract plant sap. Some are parasitic while others are predators that feed on other insects or small invertebrates. They live in a wide variety of habitats, generally terrestrial, though some species are adapted to life in or on the surface of fresh water. Hemipterans are hemimetabolous, with young nymphs that somewhat resemble adults. Many aphids are capable of parthenogenesis, producing young from unfertilised eggs; this helps them to reproduce extremely rapidly in favourable conditions.

Humans have interacted with the Hemiptera for millennia. Some species are important agricultural pests, damaging crops by the direct action of sucking sap, but also harming them indirectly by being the vectors of serious viral diseases. Other species have been used for biological control of insect pests. Hemipterans have been cultivated for the extraction of dyestuffs cochineal (also known as carmine) and for shellac. The bed bug is a persistent parasite of humans. Cicadas have been used as food, and have appeared in literature from the *Iliad* in Ancient Greece.

Diversity

Hemiptera is the largest order of hemimetabolous insects (not undergoing complete metamorphosis) containing over 75,000 named species; orders with more species all have a pupal stage, Coleoptera (370,000 described species), Lepidoptera (160,000), Diptera (100,000) and Hymenoptera (100,000). The group is very diverse. The majority of species are terrestrial, including a number of important agricultural pests, but some are found in freshwater habitats. These include the water boatmen, pond skaters, and giant water bugs.

Taxonomy and Phylogeny

The present members of the order Hemiptera (sometimes referred to as Rhynchota) were historically placed into two orders, the so-called Homoptera and Heteroptera/ Hemiptera, based on differences in wing structure and the position of the rostrum. The order is now more often divided into four or more suborders, after the "Homoptera" were established as paraphyletic. Molecular phylogenetics analysis by Song et al. (2012) supports this cladogram:

```
  ┌── Heteroptera shield bugs, assassin bugs, etc
  │
  │   ┌─ Cicadomorpha (part of Auchenorrhyncha) froghoppers, cicadas, leafhoppers, etc.
──┤   │  (35,000 species)
  │   │
  └───┤   ┌─ Fulgoromorpha (part of Auchenorrhyncha) planthoppers (over 9,000 species)
      └───┤
          └─ Sternorrhyncha aphids, whiteflies, scale insects
```

The Peloridiidae (Coleorrhyncha) were not included in Song's analysis. The suggestion that the Auchenorrhyncha are paraphyletic has been debated, and in 2012, the phylogeny was described as "contentious"; a multilocus molecular phylogenetic analysis suggested that the Auchenorrhyncha, like the Sternorrhyncha, Heteropterodea, Heteroptera, Fulgoroidea, Cicadomorpha, Membracoidea, Cercopoidea, and Cicadoidea, were all monophyletic.

Hemiptera suborders					
Suborder	**No. of Species**	**First appearance**	**Examples**	**Characteristics**	**Image**
Auchenorrhyncha	over 42,000	Lower Permian	cicadas, leafhoppers, treehoppers, planthoppers, froghoppers	plant-sucking bugs; many can jump; many make calls, some loud	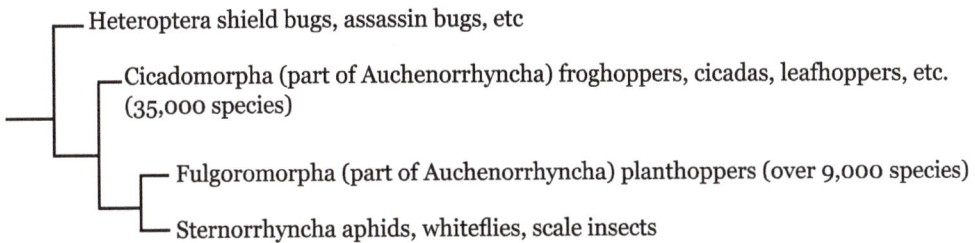

Sternorrhyncha	12,500	Upper Permian	aphids, whiteflies, scale insects	plant-sucking bugs, some major horticultural pests; most are sedentary or fully sessile	
Coleorrhyncha	fewer than 30	Lower Jurassic	moss bugs (Pelori- diidae)	evolved in the south- ern palaeo-continent of Gondwana	
Heteroptera	25,000	Triassic	shield bugs, seed bugs, assassin bugs, flower bugs, sweetpotato bugs, water bugs	larger bugs, often pred- atory	

The closest relatives of hemipterans are the thrips and lice, which collectively form the "hemipteroid assemblage" within the Exopterygota.

Fossil planthopper from the Early Cretaceous Crato Formation of Brazil, c. 116 mya

The fossil record of hemipterans goes back to the Carboniferous (Moscovian). The old- est fossils are of the Archescytinidae from the Lower Permian and are thought to be basal to the Auchenorrhyncha. Fulguromorpha and Cicadomorpha appear in the Up- per Permian, as do Sternorrhyncha of the Psylloidea and Aleurodoidea. Aphids and Coccoids appear in the Triassic. The Coleorrhyncha extend back to the Lower Jurassic. The Heteroptera first appeared in the Triassic.

Biology

Mouthparts

The defining feature of hemipterans is their "beak" in which the modified mandibles and maxillae form a "stylet" which is sheathed within a modified labium. The stylet is capable of piercing tissues and sucking liquids, while the labium supports it. The stylet contains a channel for the outward movement of saliva and another for the inward movement of liquid food. A salivary pump drives saliva into the prey; a cibarial pump extracts liquid from the prey. Both pumps are powered by substantial dilator muscles in the head. The beak is usually folded under the body when not in use. The diet is typically plant sap, but some hemipterans such as assassin bugs are blood-suckers, and a few are predators.

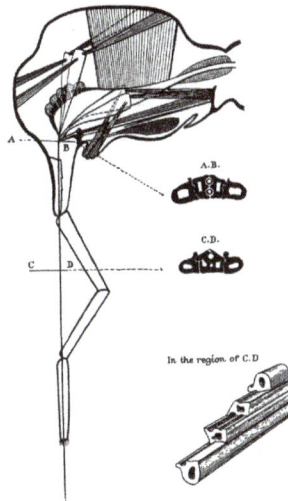

Diagram of *Hemipterous* mouth-parts. Shows the appearance, if it could be seen in perspective, with successive layers removed.

Hemipteran mouthparts are distinctive, with mandibles and maxillae modified to form a piercing "stylet" sheathed within a modified labium.

Both herbivorous and predatory hemipterans inject enzymes to begin digestion extra-orally (before the food is taken into the body). These enzymes include amylase to hydrolyse starch, polygalacturonase to weaken the tough cell walls of plants, and proteinases to break down proteins.

Although the Hemiptera vary widely in their overall form, their mouthparts form a distinctive "rostrum". Other insect orders with mouthparts modified into anything like the rostrum and stylets of the Hemiptera include some Phthiraptera, but for other reasons they generally are easy to recognize as non-hemipteran. Similarly, the mouthparts of Siphonaptera, some Diptera and Thysanoptera superficially resemble the rostrum of the Hemiptera, but on closer inspection the differences are considerable. Aside from the mouthparts, various other insects can be confused with Hemiptera, but they all have biting mandibles and maxillae instead of the rostrum. Examples include cockroaches and psocids, both of which have longer, many-segmented antennae, and some beetles, but these have fully hardened forewings which do not overlap.

Wing Structure

The forewings of Hemiptera are either entirely membranous, as in the Sternorrhyncha and Auchenorrhyncha, or partially hardened, as in most Heteroptera. Referring to the forewings of many heteropterans which are hardened near the base, but membranous at the ends. Wings modified in this manner are termed *hemelytra* (singular: *hemely-tron*), by analogy with the completely hardened elytra of beetles, and occur only in the suborder Heteroptera. In all suborders, the hindwings – if present at all – are entirely membranous and usually shorter than the forewings. The forewings may be held "roof-wise" over the body (typical of Sternorrhyncha and Auchenorrhyncha), or held flat on the back, with the ends overlapping (typical of Heteroptera). The antennae in Hemiptera typically consist of four or five segments, although they can still be quite long, and the tarsi of the legs have two or three segments.

Sound Production

Many hemipterans can produce sound for communication. The "song" of male cicadas, the loudest of any insect, is produced by tymbal organs on the underside of the abdomen, and is used to attract mates. The tymbals are drumlike disks of cuticle, which are clicked in and out repeatedly, making a sound in the same way as popping the metal lid of a jam jar in and out.

Stridulatory sounds are produced among the aquatic Corixidae and Notonectidae (backswimmers) using tibial combs rubbed across rostral ridges.

Life Cycle

An ant-mimicking predatory bug Myrmecoris gracilis

Hemipterans are hemimetabolous, meaning that they do not undergo metamorphosis, the complete change of form between a larval phase and an adult phase. Instead, their young are called nymphs, and resemble the adults to a greater or less degree. The nymphs moult several times as they grow, and each instar resembles the adult more than the previous one. Wing buds grow in later stage nymphs; the final transformation involves little more than the development of functional wings (if they are present at all) and functioning sexual organs, with no intervening pupal stage as in holometabolous insects.

Parthenogenesis and Vivipary

Aphid giving birth to live female young

Many aphids are parthenogenetic during part of the life cycle, such that females can produce unfertilized eggs, which are clones of their mother. All such young are female (thelytoky), so 100% of the population at these times can produce more offspring. Many species of aphid are also viviparous: the young are born live rather than laid as eggs. These adaptations enable aphids to reproduce extremely rapidly when conditions are suitable.

Locomotion

Hemipterans make use of a variety of modes of locomotion including swimming, skating on a water surface and jumping, as well as walking and flying like other insects.

Swimming and Skating

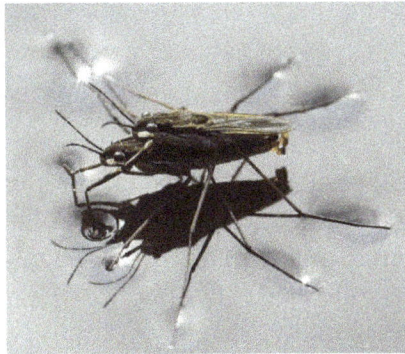

Pondskaters are adapted to use surface tension to keep above a freshwater surface.

Several families of Heteroptera are *water bugs*, adapted to an aquatic lifestyle, such as the water boatmen (Corixidae), water scorpions (Nepidae), and backswimmers (Notonectidae). They are mostly predatory, and have legs adapted as paddles to help the animal move through the water. The pondskaters or water striders (Gerridae) are also associated with water, but use the surface tension of standing water to keep them above the surface; they include the sea skaters in the genus *Halobates*, the only truly marine group of insects.

Marangoni Propulsion

Adult and nymph Microvelia water bugs using Marangoni propulsion

Marangoni effect propulsion exploits the change in surface tension when a soap-like surfactant is released on to a water surface, in the same way that a toy soap boat propels itself. Water bugs in the genus *Microvelia* (Veliidae) can travel at up to 17 cm/s, twice as fast as they can walk, by this means.

Flight

Flight is well developed in the Hemiptera although mostly used for short distance movement and dispersal. Wing development is sometimes related to environmental conditions. In aphids, both winged and wingless forms occur with winged forms produced in greater numbers when food resources are depleted. Aphids and whiteflies can sometimes be transported very long distances by atmospheric updrafts and high altitude winds.

Jumping

Many Auchenorrhyncha including representatives of the cicadas, leafhoppers, treehoppers, planthoppers, and froghoppers are adapted for jumping (saltation). Treehoppers, for example, jump by rapidly depressing their hind legs. Before jumping, the hind legs are raised and the femora are pressed tightly into curved indentations in the coxae. Treehoppers can attain a take-off velocity of up to 2.7 metres per second and an acceleration of up to 250 g. The instantaneous power output is much greater than that of normal muscle, implying that energy is stored and released to catapult the insect into the air. Cicadas, which are much larger, extend their hind legs for a jump in under a millisecond, again implying elastic storage of energy for sudden release.

Sedentary Lifestyles

In contrast, most Sternorrhyncha females are sedentary or completely sessile, attached to their host plants by their thin feeding stylets which cannot be taken out of the plant quickly.

Ecological Roles

Feeding Modes

Herbivores

Leaf galls formed by plant lice (Psyllidae), Chamaesyce celastroides var. stokesii

Most hemipterans are phytophagous, using their sucking and piercing mouthparts to feed on plant sap. These include cicadas, leafhoppers, treehoppers, planthoppers, froghoppers, aphids, whiteflies, scale insects, and some other groups. Some are mono-phages, being host specific and only found on one plant taxon, others are oligophages, feeding on a few plant groups, while others again are less discriminating polyphages and feed on many species of plant. The relationship between hemipterans and plants appears to be ancient, with piercing and sucking of plants evident in the Early Devoni-an period.

Hemipterans can dramatically cut the mass of affected plants, especially in major out-breaks. They sometimes also change the mix of plants by predation on seeds or feeding on roots of certain species. Some sap-suckers move from one host to another at differ-ent times of year. Many aphids spend the winter as eggs on a woody host plant and the summer as parthogenetically reproducing females on a herbaceous plant.

A twig wilting bug (Coreidae) piercing and sucking sap from a *Zinnia*

Phloem sap, which has a higher concentration of sugars and nitrogen, is under positive pressure unlike the more dilute xylem sap. Most of the Sternorrhyncha and a number of Auchenorrhynchan groups feed on phloem. Phloem feeding is common in the Fulgoromorpha, most Cicadellidae and in the Heteroptera. The Typhlocybine Cicadellids specialize in feeding on non-vascular mesophyll tissue of leaves, which is more nutritious than the leaf epidermis. Most Heteroptera also feed on mesophyll tissue where they are more likely to encounter defensive secondary plant metabolites which often leads to the evolution of host specificity. Obligate xylem feeding is a special habit that is found in the Auchenorrhyncha among Cicadoidea, Cercopoidea and in Cicadelline Cicadellids. Some phloem feeders may take to xylem sap facultatively, especially when facing dehydration. Xylem feeders tend to be polyphagous; to overcome the negative pressure of xylem requires a special cibarial pump. Phloem feeding hemiptera typically have symbiotic micro-organisms in their gut that help to convert amino acids. Phloem feeders produce honeydew from their anus. A variety of organisms that feed on honeydew form symbiotic associations with phloem-feeders. Phloem sap is a sugary liquid low in amino acids, so insects have to process large quantities to meet their nutritional requirements. Xylem sap is even lower in amino acids and contains monosaccharides rather than sucrose, as well as organic acids and minerals. No digestion is required (except for the hydrolysis of sucrose) and 90% of the nutrients in the xylem sap can be utilised. Some phloem sap feeders selectively mix phloem and xylem sap to control the osmotic potential of the liquid consumed. A striking adaptation to a very dilute diet is found in many hemipterans: a filter chamber, a part of the gut looped back on itself as a countercurrent exchanger, which permits nutrients to be separated from excess water. The residue, mostly water with sugars and amino acids, is quickly excreted as sticky "honey dew", notably from aphids but also from other Auchenorrhycha and Sternorrhyncha.

Some Sternorrhyncha including Psyllids and some aphids are gall formers. These sap-sucking hemipterans inject fluids containing plant hormones into the plant tissues inducing the production of tissue that covers to protects the insect and also act as sinks for nutrition that they feed on. The hackleberry gall psyllid for example, causes a woody gall on the leaf petioles of the hackleberry tree it infests, and the nymph of another psyllid produces a protective lerp out of hardened honeydew.

Predators

Most other hemipterans are predatory, feeding on other insects, or even small vertebrates. This is true of many aquatic species which are predatory, either as nymphs or adults. The predatory shield bug for example stabs caterpillars with its beak and sucks out the body fluids. The saliva of predatory heteropterans contains digestive enzymes such as proteinase and phospholipase, and in some species also amylase. The mouthparts of these insects are adapted for predation. There are toothed stylets on the mandibles able to cut into and abrade tissues of their prey. There are further stylets on the maxillae, adapted as tubular canals to inject saliva and to extract the pre-digested and liquified contents of the prey.

Some species attack pest insects and are used in biological control. One of these is the spined soldier bug (*Podisus maculiventris*) that sucks body fluids from larvae of the Colorado beetle and the Mexican bean beetle.

Haematophagic "Parasites"

A few hemipterans are haematophagic (often described as "parasites"), feeding on the blood of larger animals. These include bedbugs and the triatomine kissing bugs of the assassin bug family Reduviidae, which can transmit the dangerous *Chagas disease*. The first known hemipteran to feed in this way on vertebrates was the extinct assassin bug *Triatoma dominicana* found fossilized in amber and dating back about twenty million years. Faecal pellets fossilised beside it show that it transmitted a disease-causing *Trypanosoma* and the amber included hairs of the likely host, a bat.

As Symbionts

Leafhoppers protected by meat ants

Some species of ant protect and farm aphids (Sternorrhyncha) and other sap-sucking hemipterans, gathering and eating the honeydew that these hemipterans secrete. The relationship is symbiotic, as both ant and aphid benefit. Ants such as the yellow anthill ant, *Lasius flavus*, breed aphids of at least four species, *Geoica utricularia*, *Tetraneura ulmi*, *Forda marginata* and *Forda formicaria*, taking eggs with them when they found a new colony; in return, these aphids are obligately associated with the ant, breeding mainly or wholly asexually inside anthills. Ants may also protect the plant bugs from their natural enemies, removing the eggs of predatory beetles and preventing access by parasitic wasps.

Some leafhoppers (Auchenorrhyncha) are similarly "milked" by ants. In the Corcovado rain forest of Costa Rica, wasps compete with ants to protect and milk leafhoppers; the leafhoppers preferentially gave more honeydew, more often, to the wasps, which were larger and may have offered better protection.

As Prey: Defences Against Predators and Parasites

Masked hunter nymph camouflaged with sand grains

Hemiptera form prey to predators including vertebrates, such as birds, and other invertebrates such as ladybirds. In response, hemipterans have evolved antipredator adaptations. *Ranatra* may feign death (thanatosis). Others such as *Carpocoris purpureipennis* secrete toxic fluids to ward off arthropod predators; some Pentatomidae such as *Dolycoris* are able to direct these fluids at an attacker. Toxic cardenolide compounds are accumulated by the heteropteran *Oncopeltus fasciatus* when it consumes milkweeds, while the coreid stinkbug *Amorbus rubiginosus* acquires 2-hexenal from its food plant, *Eucalyptus*. Some long-legged bugs mimic twigs, rocking to and fro to simulate the motion of a plant part in the wind. The nymph of the Masked hunter bug camouflages itself with sand grains, using its hind legs and tarsal fan to form a double layer of grains, coarser on the outside. The Amazon rain forest cicada *Hemisciera maculipennis* display bright red deimatic flash coloration on their hindwings when threatened; the sudden contrast helps to startle predators, giving the cicadas time to escape. The coloured patch on the hindwing is concealed at rest by an olive green patch of the same size on the forewing, enabling the insect to switch rapidly from cryptic to deimatic behaviour.

Firebugs, *Pyrrhocoris apterus*, protect themselves from predators with bright aposematic warning coloration, and by aggregating in a group.

Some hemipterans such as firebugs have bold aposematic warning coloration, often red and black, which appear to deter passerine birds. Many hemipterans including aphids, scale insects and especially the planthoppers secrete wax to protect themselves from threats such as fungi, parasitoidal insects and predators, as well as abiotic factors like desiccation. Hard waxy coverings are especially important in the sedentary Sternor-rhyncha such as scale insects, which have no means of escape from predators; other Sternorrhyncha evade detection and attack by creating and living inside plant galls. Nymphal Cicadoidea and Cercopoidea have glands attached to the Malpighian tubules in their proximal segment that produce mucopolysaccharides, which form the froth around spittlebugs, offering a measure of protection.

Parental care is found in many species of Hemiptera especially in members of the Membracidae and numerous Heteroptera. In many species of shield bug, females stand guard over their egg clusters to protect them from egg parasitoids and predators. In the aquatic Belostomatidae, females lay their eggs on the back of the male which guards the eggs. Protection provided by ants is common in the Auchenorrhyncha.

Interaction with Humans

Colony of cottony cushion scale, a pest of citrus fruits

As Pests

Although many species of Hemiptera are significant pests of crops and garden plants, including many species of aphid and scale insects, other species are harmless. The damage done is often not so much the deprivation of the plant of its sap, but the fact that they transmit serious viral diseases between plants. They often produce copious amounts of honeydew which encourages the growth of sooty mould. Significant pests include the cottony cushion scale, a pest of citrus fruit trees, the green peach aphid and other aphids which attack crops worldwide and transmit diseases, and jumping plant lice which are often host plant-specific and transmit diseases.

For Pest Control

Members of the families Reduviidae, Phymatidae and Nabidae are obligate predators. Some predatory species are used in biological pest control; these include various na-bids, and even some members of families that are primarily phytophagous, such as the genus *Geocoris* in the family Lygaeidae. Other hemipterans are omnivores, alternating between a plant-based and an animal-based diet. For example, *Dicyphus hesperus* is used to control whitefly on tomatoes but also sucks sap, and if deprived of plant tissues will die even if in the presence of whiteflies.

Insect Products

Cochineal scale insects being collected from a prickly pear in Central America. Illustration by José Antonio de Alzate y Ramírez, 1777

Other hemipterans have positive uses for humans, such as in the production of the dye-stuff carmine (cochineal). The FDA has created guidelines for how to declare when it has been added to a product. The scale insect *Dactylopius coccus* produces the brilliant red-coloured carminic acid to deter predators. Up to 100,000 scale insects need to be collected and processed to make a kilogram (2.2 lbs) of cochineal dye. A similar number of lac bugs are needed to make a kilogram of shellac, a brush-on colourant and wood finish. Additional uses of this traditional product include the waxing of citrus fruits to extend their shelf-life, and the coating of pills to moisture-proof them, provide slow-release or mask the taste of bitter ingredients.

As Human Parasites and Disease Vectors

Bed bug nymph, Cimex lectularius, engorged with human blood

Chagas disease is a modern-day tropical disease caused by *Trypanosoma cruzi* and transmitted by kissing bugs, so-called because they suck human blood from around the lips while a person sleeps.

The bed bug, *Cimex lectularius*, is an external parasite of humans. It lives in bedding and is mainly active at night, feeding on human blood, generally without being noticed. Bed bugs mate by traumatic insemination; the male pierces the female's abdomen and injects his sperm into a secondary genital structure, the spermalege. The sperm travel in the female's blood (haemolymph) to sperm storage structures (seminal conceptacles); they are released from there to fertilise her eggs inside her ovaries.

As Food

Deep-fried cicadas, Cryptotympana atrata, in Chinese Shandong cuisine

Some larger hemipterans such as cicadas are used as food in Asian countries such as China, and they are much esteemed in Malawi and other African countries. Insects have a high protein content and good food conversion ratios, but most hemipterans are too small to be a useful component of the human diet. At least nine species of Hemiptera are eaten worldwide.

In Art and Literature

Cicadas have featured in literature since the time of Homer's *Iliad*, and as motifs in decorative art from the Chinese Shang dynasty (1766-1122 B.C.). They are described by Aristotle in his *History of Animals* and by Pliny the Elder in his *Natural History*; their mechanism of sound production is mentioned by Hesiod in his poem *Works and Days* "when the Skolymus flowers, and the tuneful *Tettix* sitting on his tree in the weary summer season pours forth from under his wings his shrill song".

In Mythology and Folklore

Among the bugs, cicadas in particular have been used as money, in folk medicine, to forecast the weather, to provide song (in China), and in folklore and myths around the world.

Threats

Large-scale cultivation of the oil palm *Elaeis guineensis* in the Amazon basin damages freshwater habitats and reduces the diversity of aquatic and semi-aquatic Heteroptera.

Climate change may be affecting the global migration of hemipterans including the potato leafhopper, *Empoasca fabae*. Warming is correlated with the severity of potato leafhopper infestation, so increased warming may worsen infestations in future.

Termite

Termites are eusocial insects that are classified at the taxonomic rank of infraorder Isoptera, or as epifamily Termitoidae within the cockroach order Blattodea. Termites were once classified in a separate order from cockroaches, but recent phylogenetic studies indicate that they evolved from close ancestors of cockroaches during the Jurassic or Triassic. However, the first termites possibly emerged during the Permian or even the Carboniferous. About 3,106 species are currently described, with a few hundred more left to be described. Although these insects are often called white ants, they are not ants.

Like ants and some bees and wasps from the separate order Hymenoptera, termites divide labour among castes consisting of sterile male and female "workers" and "soldiers". All colonies have fertile males called "kings" and one or more fertile females called "queens". Termites mostly feed on dead plant material and cellulose, generally in the form of wood, leaf litter, soil, or animal dung. Termites are major detritivores, particularly in the subtropical and tropical regions, and their recycling of wood and plant matter is of considerable ecological importance.

Termites are among the most successful groups of insects on Earth, colonising most landmasses except for Antarctica. Their colonies range in size from a few hundred individuals to enormous societies with several million individuals. Termite queens have the longest lifespan of any insect in the world, with some queens living up to 50 years. Unlike ants, which undergo a complete metamorphosis, each individual termite goes through an incomplete metamorphosis that proceeds through egg, nymph, and adult stages. Colonies are described as superorganisms because the termites form part of a self-regulating entity: the colony itself.

Termites are a delicacy in the diet of some human cultures and are used in many traditional medicines. Several hundred species are economically significant as pests that can cause serious damage to buildings, crops, or plantation forests. Some species, such as the West Indian drywood termite (*Cryptotermes brevis*), are regarded as invasive species.

Etymology

The infraorder name Isoptera is derived from the Greek words *iso* (equal) and *ptera* (winged), which refers to the nearly equal size of the fore and hind wings. "Termite" derives from the Latin and Late Latin word *termes* ("woodworm, white ant"), altered by the influence of Latin *terere* ("to rub, wear, erode") from the earlier word *tarmes*.

Termite nests were commonly known as *terminarium* or *termitaria*. In early English, termites were known as "wood ants" or "white ants". The modern term was first used in 1781.

Taxonomy and Evolution

The external appearance of the giant northern termite Mastotermes darwiniensis is suggestive of the close relationship between termites and cockroaches.

DNA analysis from 16S rRNA sequences has supported a hypothesis, originally suggested by Cleveland and colleagues in 1934, that these insects are most closely related to wood-eating cockroaches (genus *Cryptocercus*, the woodroach). This earlier conclusion had been based on the similarity of the symbiotic gut flagellates in the wood-eating cockroaches to those in certain species of termites regarded as living fossils. In the 1960s additional evidence supporting that hypothesis emerged when F. A. McKittrick noted similar morphological characteristics between some termites and *Cryptocercus* nymphs. These similarities have led some authors to propose that termites be reclassified as a single family, the Termitidae, within the order Blattodea, which contains cockroaches. Other researchers advocate the more conservative measure of retaining the termites as the Termitoidae, an epifamily within the cockroach order, which preserves the classification of termites at family level and below.

The oldest unambiguous termite fossils date to the early Cretaceous, but given the diversity of Cretaceous termites and early fossil records showing mutualism between microorganisms and these insects, they likely originated earlier in the Jurassic or Triassic. Further evidence of a Jurassic origin is the assumption that the extinct *Fruitafossor* consumed termites, judging from its morphological similarity to modern termite-eating mammals. The oldest termite nest discovered is believed to be from the Upper Cretaceous in West Texas, where the oldest known faecal pellets were also discovered.

Claims that termites emerged earlier have faced controversy. For example, F. M. Weesner indicated that the Mastotermitidae termites may go back to the Late Permian, 251 million years ago, and fossil wings that have a close resemblance to the wings of *Mastotermes* of the Mastotermitidae, the most primitive living termite, have been discovered in the Permian layers in Kansas. It is even possible that the first termites emerged during the Carboniferous. Termites are thought to be the descendants of the genus *Cryptocercus*. The folded wings of the fossil wood roach *Pycnoblattina*, arranged in a convex pattern between segments 1a and 2a, resemble those seen in *Mastotermes*, the only living insect with the same pattern. Krishna *et al.*, though, consider that all of the Paleozoic and Triassic insects tentatively classified as termites are in fact unrelated to termites and should be excluded from the Isoptera. Termites were the first social insects to evolve a caste system, evolving more than 100 million years ago.

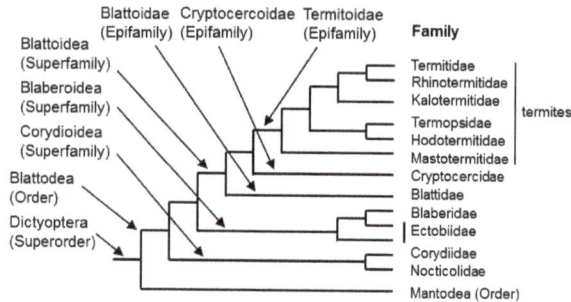

Evolutionary relationships of Blattodea, showing the placement of some termite families

Termites have long been accepted to be closely related to cockroaches and mantids, and they are classified in the same superorder (Dictyoptera). Strong evidence suggests termites are highly specialised wood-eating cockroaches. The cockroach genus *Cryptocercus* shares the strongest phylogenetical similarity with termites and is considered to be a sister-group to termites. Termites and *Cryptocercus* share similar morphological and social features: for example, most cockroaches do not exhibit social characteristics, but *Cryptocercus* takes care of its young and exhibits other social behaviour such as trophallaxis and allogrooming. The primitive giant northern termite (*Mastotermes darwiniensis*) exhibits numerous cockroach-like characteristics that are not shared with other termites, such as laying its eggs in rafts and having anal lobes on the wings. Cryptocercidae and Isoptera are united in the clade Xylophagodea. Although termites are sometimes called "white ants", they are actually not ants. Ants belong to the family Formicidae within the order Hymenoptera. The similarity of their social structure to that of termites is attributed to convergent evolution.

As of 2013, about 3,106 living and fossil termite species are recognised, classified in 12 families. The infraorder Isoptera is divided into the following clade and family groups, showing the subfamilies in their respective classification:

Distribution and Diversity

Termites are found on all continents except Antarctica. The diversity of termite species is low in North America and Europe (10 species known in Europe and 50 in North America), but is high in South America, where over 400 species are known. Of the 3,000 termite species currently classified, 1,000 are found in Africa, where mounds are extremely abundant in certain regions. Approximately 1.1 million active termite mounds can be found in the northern Kruger National Park alone. In Asia, there are 435 species of termites, which are mainly distributed in China. Within China, termite species are restricted to mild tropical and subtropical habitats south of the Yangtze River. In Australia, all ecological groups of termites (dampwood, drywood, subterranean) are endemic to the country, with over 360 classified species.

Due to their soft cuticles, termites do not inhabit cool or cold habitats. There are three ecological groups of termites: dampwood, drywood and subterranean. Dampwood termites are found only in coniferous forests, and drywood termites are found in hardwood forests; subterranean termites live in widely diverse areas. One species in the drywood group is the West Indian drywood termite *(Cryptotermes brevis)*, which is an invasive species in Australia.

Diversity of Isoptera by continent:						
	Asia	Africa	North America	South America	Europe	Australia
Estimated number of species	435	1,000	50	400	10	360

Description

Close-up view of a worker's head

Termites are usually small, measuring between 4 to 15 millimetres (0.16 to 0.59 in) in length. The largest of all extant termites are the queens of the species *Macrotermes bellicosus*, measuring up to over 10 centimetres (4 in) in length. Another giant termite, the extinct *Gyatermes styriensis*, flourished in Austria during the Miocene and had a wingspan of 76 millimetres (3.0 in) and a body length of 25 millimetres (0.98 in).

Most worker and soldier termites are completely blind as they do not have a pair of eyes. However, some species, such as *Hodotermes mossambicus*, have compound eyes which they use for orientation and to distinguish sunlight from moonlight. The alates have eyes along with lateral ocelli. Lateral ocelli, however, are not found in all termites. Like other insects, termites have a small tongue-shaped labrum and a clypeus; the clypeus is divided into a postclypeus and anteclypeus. Termite antennae have a number of functions such as the sensing of touch, taste, odours (including pheromones), heat and vibration. The three basic segments of a termite antenna include a scape, a pedicel (typically shorter than the scape), and the flagellum (all segments beyond the scape and pedicel). The mouth parts contain a maxillae, a labium, and a set of mandibles. The maxillae and labium have palps that help termites sense food and handling.

Consistent with all insects, the anatomy of the termite thorax consists of three segments: the prothorax, the mesothorax and the metathorax. Each segment contains a pair of two legs. On alates, the wings are located at the mesothorax and metathorax. The mesothorax and metathorax have well-developed exoskeletal plates; the prothorax has smaller plates.

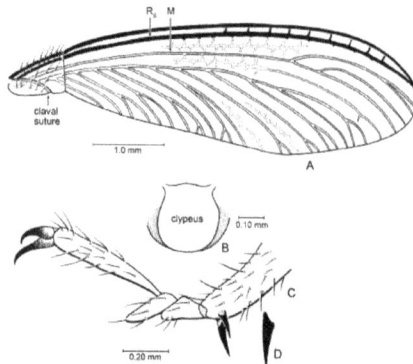

Diagram showing a wing, along with the clypeus and leg

Termites have a ten-segmented abdomen with two plates, the tergites and the sternites. There are ten tergites, of which nine are wide and one is elongated. The reproductive organs are similar to those in cockroaches but are more simplified. For example, the intromittent organ is not present in male alates, and the sperm is either immotile or aflagellate. However, Mastotermitidae termites have multiflagellate sperm with limited motility. The genitals in females are also simplified. Unlike in other termites, Mastotermitidae females have an ovipositor, a feature strikingly similar to that in female cockroaches.

The non-reproductive castes of termites are wingless and rely exclusively on their six legs for locomotion. The alates fly only for a brief amount of time, so they also rely on their legs. The appearance of the legs is similar in each caste, but the soldiers have larger and heavier legs. The structure of the legs is consistent with other insects: the parts of a leg include a coxa, trochanter, femur, tibia and the tarsus. The number of tibial spurs on an individual's leg varies. Some species of termite have an arolium, located between the claws, which is present in species that climb on smooth surfaces but is absent in most termites.

Unlike in ants, the hind-wings and fore-wings are of equal length. Most of the time, the alates are poor flyers; their technique is to launch themselves in the air and fly in a random direction. Studies show that in comparison to larger termites, smaller termites cannot fly long distances. When a termite is in flight, its wings remain at a right angle, and when the termite is at rest, its wings remain parallel to the body.

Caste System

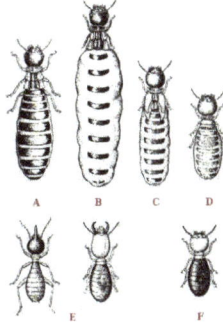

Caste system of termites
A — King
B — Queen
C — Secondary queen
D — Tertiary queen
E — Soldiers
F — Worker

Worker termites undertake the most labour within the colony, being responsible for foraging, food storage, and brood and nest maintenance. Workers are tasked with the digestion of cellulose in food and are thus the most likely caste to be found in infested wood. The process of worker termites feeding other nestmates is known as trophallaxis. Trophallaxis is an effective nutritional tactic to convert and recycle nitrogenous components. It frees the parents from feeding all but the first generation of offspring, allowing for the group to grow much larger and ensuring that the necessary gut symbionts are transferred from one generation to another. Some termite species do not have a true worker caste, instead relying on nymphs that perform the same work without differentiating as a separate caste.

The soldier caste has anatomical and behavioural specialisations, and their sole purpose is to defend the colony. Many soldiers have large heads with highly modified powerful jaws so enlarged they cannot feed themselves. Instead, like juveniles, they are fed by workers. Fontanelles, simple holes in the forehead that exude defensive secretions, are a feature of the family Rhinotermitidae. Many species are readily identified using the characteristics of the soldiers' larger and darker head and large mandibles. Among certain termites, soldiers may use their globular (phragmotic) heads to block their narrow tunnels. Different sorts of soldiers include minor and major soldiers, and nasutes, which have a horn-like nozzle frontal projection (a nasus). These unique soldiers are able to spray noxious, sticky secretions containing diterpenes at their enemies. Nitrogen fixation plays an important role in nasute nutrition.

The reproductive caste of a mature colony includes a fertile female and male, known as the queen and king. The queen of the colony is responsible for egg production for the colony. Unlike in ants, the king mates with her for life. In some species, the abdomen of the queen swells up dramatically to increase fecundity, a characteristic known as physogastrism. Depending on the species, the queen will start producing reproductive winged alates at a certain time of the year, and huge swarms emerge from the colony when nuptial flight begins. These swarms attract a wide variety of predators.

Life Cycle

A young termite nymph. Nymphs first moult into workers, but others may further moult to become soldiers or alates.

Termites are often compared with the social Hymenoptera (ants and various species of bees and wasps), but their differing evolutionary origins result in major differences in life cycle. In the eusocial Hymenoptera, the workers are exclusively female: males (drones) are haploid and develop from unfertilised eggs, while females (both workers and the queen) are diploid and develop from fertilised eggs. In contrast, worker termites, which constitute the majority in a colony, are diploid individuals of both sexes and develop from fertilised eggs. Depending on species, male and female workers may have different roles in a termite colony.

The life cycle of a termite begins with an egg, but is different from that of a bee or ant in that it goes through a developmental process called incomplete metamorphosis, with egg, nymph and adult stages. Nymphs resemble small adults, and go through a series of moults as they grow. In some species, eggs go through four moulting stages and nymphs go through three. Nymphs first moult into workers, and then some workers go through further moulting and become soldiers or alates; workers become alates only by moulting into alate nymphs.

The development of nymphs into adults can take months; the time period depends on food availability, temperature, and the general population of the colony. Since nymphs are unable to feed themselves, workers must feed them, but workers also take part in the social life of the colony and have certain other tasks to accomplish such as foraging, building or maintaining the nest or tending to the queen. Pheromones regulate the caste system in termite colonies, preventing all but a very few of the termites from becoming fertile queens.

Reproduction

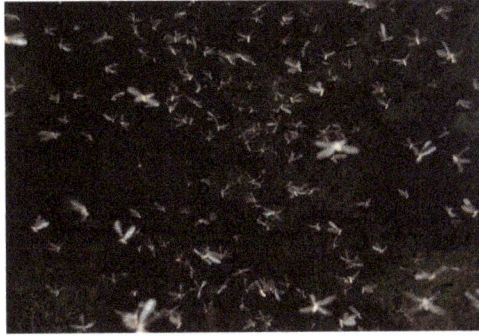

Alates swarming during nuptial flight after rain

Termite alates only leave the colony when a nuptial flight takes place. Alate males and females will pair up together and then land in search of a suitable place for a colony. A termite king and queen will not mate until they find such a spot. When they do, they excavate a chamber big enough for both, close up the entrance and proceed to mate. After mating, the pair will never go outside and will spend the rest of their lives in the nest. Nuptial flight time varies in each species. For example, alates in certain species emerge during the day in summer while others emerge during the winter. The nuptial flight may also begin at dusk, when the alates swarm around areas with lots of lights. The time when nuptial flight begins depends on the environmental conditions, the time of day, moisture, wind speed and precipitation. The number of termites in a colony also varies, with the larger species typically having 100–1,000 individuals. However, some termite colonies, including those with large individuals, can number in the millions.

The queen will only lay 10–20 eggs in the very early stages of the colony, but will lay as many as 1,000 a day when the colony is several years old. At maturity, a primary queen has a great capacity to lay eggs. In some species, the mature queen has a greatly distended abdomen and may produce 40,000 eggs a day. The two mature ovaries may have some 2,000 ovarioles each. The abdomen increases the queen's body length to several times more than before mating and reduces her ability to move freely; attendant workers provide assistance.

Egg grooming behaviour of *Reticulitermes speratus* workers in a nursery cell

The king grows only slightly larger after initial mating and continues to mate with the queen for life (a termite queen can live up to 50 years). This is very different from ant colonies, in which a queen mates once with the male(s) and stores the gametes for life, as the male ants die shortly after mating. If a queen is absent, a termite king will produce pheromones which encourage the development of replacement termite queens. As the queen and king are monogamous, sperm competition does not occur.

Termites going through incomplete metamorphosis on the path to becoming alates form a subcaste in certain species of termite, functioning as potential supplementary reproductives. These supplementary reproductives only mature into primary reproductives upon the death of a king or queen, or when the primary reproductives are separated from the colony. Supplementaries have the ability to replace a dead primary reproductive, and there may also be more than a single supplementary within a colony. Some queens have the ability to switch from sexual reproduction to asexual reproduction. Studies show that while termite queens mate with the king to produce colony workers, the queens reproduce their replacements (neotenic queens) parthenogenetically.

Behaviour and Ecology

Diet

Termite faecal pellets

Termites are detritivores, consuming dead plants at any level of decomposition. They also play a vital role in the ecosystem by recycling waste material such as dead wood, faeces and plants. Many species eat cellulose, having a specialised midgut that breaks down the fibre. Termites are considered to be a major source (11%) of atmospheric methane, one of the prime greenhouse gases, produced from the breakdown of cellulose. Termites rely primarily upon symbiotic protozoa (metamonads) and other microbes such as flagellate protists in their guts to digest the cellulose for them, allowing them to absorb the end products for their own use. Gut protozoa, such as *Trichonympha*, in turn, rely on symbiotic bacteria embedded on their surfaces to produce some of the necessary digestive enzymes. Most higher termites, especially in the family Termitidae, can produce their own cellulase enzymes, but they rely primarily upon the bacteria. The flagellates have been lost in Termitidae. Scientists' understanding of the relationship between the termite digestive tract and the microbial endosymbionts is still rudimentary; what is true in all termite species, however, is that the workers feed the other members of the colony with substances derived from the digestion of plant material, either from the mouth or anus. Judging from closely related bacterial species, it is strongly presumed that the termites' and cockroach's gut microbiota derives from their dictyopteran ancestors.

Certain species such as *Gnathamitermes tubiformans* have seasonal food habits. For example, they may preferentially consume Red three-awn (*Aristida longiseta*) during the summer, Buffalograss (*Buchloe dactyloides*) from May to August, and blue grama *Bouteloua gracilis* during spring, summer and autumn. Colonies of *G. tubiformans* consume less food in spring than they do during autumn when their feeding activity is high.

Various woods differ in their susceptibility to termite attack; the differences are attributed to such factors as moisture content, hardness, and resin and lignin content. In one study, the drywood termite *Cryptotermes brevis* strongly preferred poplar and maple woods to other woods that were generally rejected by the termite colony. These preferences may in part have represented conditioned or learned behaviour.

Some species of termite practice fungiculture. They maintain a "garden" of specialised fungi of genus *Termitomyces*, which are nourished by the excrement of the insects. When the fungi are eaten, their spores pass undamaged through the intestines of the termites to complete the cycle by germinating in the fresh faecal pellets.

Depending on their feeding habits, termites are placed into two groups: the lower termites and higher termites. The lower termites predominately feed on wood. As wood is difficult to digest, termites prefer to consume fungus-infected wood because it is easier to digest and the fungi are high in protein. Meanwhile, the higher termites consume a wide variety of materials, including faeces, humus, grass, leaves and roots. The gut in the lower termites contains many species of bacteria along with protozoa, while the higher termites only have a few species of bacteria with no protozoa.

Predators

Crab spider with a captured alate

Termites are consumed by a wide variety of predators. One species alone, *Hodotermes mossambicus*, was found in the stomach contents of 65 birds and 19 mammals. Arthropods and reptiles such as bees, centipedes, cockroaches, crickets, dragonflies, frogs, lizards, scorpions, spiders, and toads consume these insects, while two spiders in the family Ammoxenidae are specialist termite predators. Other predators include aardvarks, aardwolves, anteaters, bats, bears, bilbies, many birds, echidnas, foxes, galagos, numbats, mice and pangolins. The aardwolf is an insectivorous mammal that primarily feeds on termites; it locates its food by sound and also by detecting the scent secreted by the soldiers; a single aardwolf is capable of consuming thousands of termites in a single night by using its long, sticky tongue. Sloth bears break open mounds to consume the nestmates, while chimpanzees have developed tools to "fish" termites from their nest. Wear pattern analysis of bone tools used by the early hominin *Paranthropus robustus* suggests that they used these tools to dig into termite mounds.

A Matabele ant (Megaponera analis) kills a Macrotermes bellicosus termite soldier during a raid.

Among all predators, ants are the greatest enemy to termites. Some ant genera are specialist predators of termites. For example, *Megaponera* is a strictly termite-eating (termitophagous) genus that perform raiding activities, some lasting several hours. *Paltothyreus tarsatus* is another termite-raiding species, with each individual stacking as many termites as possible in its mandibles before returning home, all the while recruiting additional nestmates to the raiding site through chemical trails. The Malaysian basicerotine ant *Eurhopalothrix heliscata* uses a different strategy of termite hunting by pressing themselves into tight spaces, as they hunt through rotting wood housing termite colonies. Once inside, the ants seize their prey by using their short but sharp mandibles. *Tetramorium uelense* is a specialised predator species that feeds on small termites. A scout will recruit 10–30 workers to an area where termites are present, killing them by immobilising them with their stinger. *Centromyrmex* and *Iridomyrmex* colonies sometimes nest in termite mounds, and so the termites are preyed on by these ants. No evidence for any kind of relationship (other than a predatory one) is known. Other ants, including *Acanthostichus, Camponotus, Crematogaster, Cylindromyrmex, Leptogenys, Odontomachus, Ophthalmopone, Pachycondyla, Rhytidoponera, Solenopsis* and *Wasmannia*, also prey on termites. In contrast to all these ant species, and despite their enormous diversity of prey, *Dorylus* ants rarely consume termites.

Ants are not the only invertebrates that perform raids. Many sphecoid wasps and several species including *Polybia Lepeletier* and *Angiopolybia Araujo* are known to raid termite mounds during the termites' nuptial flight.

Parasites, Pathogens and Viruses

Termites are less likely to be attacked by parasites than bees, wasps and ants, as they are usually well protected in their mounds. Nevertheless, termites are infected by a variety of parasites. Some of these include dipteran flies, *Pyemotes* mites, and a large number of nematode parasites. Most nematode parasites are in the order Rhabditida; others are in the genus *Mermis*, *Diplogaster aerivora* and *Harteria gallinarum*. Under imminent threat of an attack by parasites, a colony may migrate to a new location. Fungi pathogens such as such as *Aspergillus nomius* and *Metarhizium anisopliae* are, however, major threats to a termite colony as they are not host-specific and may infect large portions of the colony; transmission usually occurs via direct physical contact. *M. anispliae* is known to weaken the termite immune system. Infection with *A. nomius* only occurs when a colony is under great stress. Inquilinism between two termite species does not occur in the termite world.

Termites are infected by viruses including Entomopoxvirinae and the Nuclear Polyhedrosis Virus.

Locomotion and Foraging

Because the worker and soldier castes lack wings and thus never fly, and the reproductives use their wings for just a brief amount of time, termites predominantly rely upon their legs to move about.

Foraging behaviour depends on the type of termite. For example, certain species feed on the wood structures they inhabit, and others harvest food that is near the nest. Most workers are rarely found out in the open, and do not forage unprotected; they rely on sheeting and runways to protect them from predators. Subterranean termites construct tunnels and galleries to look for food, and workers who manage to find food sources recruit additional nestmates by depositing a phagostimulant pheromone that attracts workers. Foraging workers use semiochemicals to communicate with each other, and workers who begin to forage outside of their nest release trail pheromones from their sternal glands. In one species, *Nasutitermes costalis*, there are three phases in a foraging expedition: first, soldiers scout an area. When they find a food source, they communicate to other soldiers and a small force of workers starts to emerge. In the second phase, workers appear in large numbers at the site. The third phase is marked by a decrease in the number of soldiers present and an increase in the number of workers. Isolated termite workers may engage in Lévy flight behaviour as an optimised strategy for finding their nestmates or foraging for food.

Competition

Competition between two colonies always results in agonistic behaviour towards each other, resulting in fights. These fights can cause mortality on both sides and, in some cases, the gain or loss of territory. "Cemetery pits" may be present, where the bodies of dead termites are buried.

Studies show that when termites encounter each other in foraging areas, some of the termites deliberately block passages to prevent other termites from entering. Dead termites from other colonies found in exploratory tunnels leads to the isolation of the area and thus the need to construct new tunnels. Conflict between two competitors does not always occur. For example, though they might block each other's passages, colonies of *Macrotermes bellicosus* and *Macrotermes subhyalinus* are not always aggressive towards each other. Suicide cramming is known in *Coptotermes formosanus*. Since *C. formosanus* colonies may get into physical conflict, some termites will tightly squeeze into foraging tunnels and die, successfully blocking the tunnel and ending all agonistic activities.

Among the reproductive caste, neotenic queens may compete with each other to become the dominant queen when there are no primary reproductives. This struggle among the queens leads to the elimination of all but a single queen, which, with the king, will take over the colony.

Ants and termites may compete with each other for nesting space. In particular, ants that prey on termites usually have a negative impact on arboreal nesting species.

Communication

Hordes of Nasutitermes on a march for food, following, and leaving, trail pheromones

Most termites are blind, so communication primarily occurs through chemical, mechanical and pheromonal cues. These methods of communication are used in a variety of activities, including foraging, locating reproductives, construction of nests, recognition of nestmates, nuptial flight, locating and fighting enemies, and defending the nests. The most common way of communicating is through antennation. A number of pheromones are known, including contact pheromones (which are transmitted when workers are engaged in trophallaxis or grooming) and alarm, trail and sex pheromones. The alarm pheromone and other defensive chemicals are secreted from the frontal gland. Trail pheromones are secreted from the sternal gland, and sex pheromones derive from two glandular sources: the sternal and tergal glands. When termites go out to look for food, they forage in columns along the ground through vegetation. A trail can be identified by the faecal deposits or runways that are covered by objects. Workers leave pheromones on these trails, which are detected by other nestmates through olfactory receptors. Termites can also communicate through mechanical cues, vibrations, and physical contact. These signals are frequently used for alarm communication or for evaluating a food source.

When termites construct their nests, they use predominantly indirect communication. No single termite would be in charge of any particular construction project. Individual termites react rather than think, but at a group level, they exhibit a sort of collective cognition. Specific structures or other objects such as pellets of soil or pillars cause termites to start building. The termite adds these objects onto existing structures, and such behaviour encourages building behaviour in other workers. The result is a self-organised process whereby the information that directs termite activity results from changes in the environment rather than from direct contact among individuals.

Termites can distinguish nestmates and non-nestmates through chemical communication and gut symbionts: chemicals consisting of hydrocarbons released from the cuticle allow the recognition of alien termite species. Each colony has its own distinct odour. This odour is a result of genetic and environmental factors such as the termites' diet and the composition of the bacteria within the termites' intestines.

Defence

Termites rush to a damaged area of the nest.

Termites rely on alarm communication to defend a colony. Alarm pheromones can be released when the nest has been breached or is being attacked by enemies or potential pathogens. Termites always avoid nestmates infected with *Metarhizium anisopliae* spores, through vibrational signals released by infected nestmates. Other methods of defence include intense jerking and secretion of fluids from the frontal gland and defecating faeces containing alarm pheromones.

In some species, some soldiers block tunnels to prevent their enemies from entering the nest, and they may deliberately rupture themselves as an act of defence. In cases where the intrusion is coming from a breach that is larger than the soldier's head, defence requires a special formations where soldiers form a phalanx-like formation around the breach and bite at intruders. If an invasion carried out by *Megaponera analis* is successful, an entire colony may be destroyed, although this scenario is rare.

To termites, any breach of their tunnels or nests is a cause for alarm. When termites detect a potential breach, the soldiers will usually bang their heads apparently to attract other soldiers for defence and to recruit additional workers to repair any breach. Additionally, an alarmed termite will bump into other termites which causes them to be alarmed and to leave pheromone trails to the disturbed area, which is also a way to recruit extra workers.

Nasute termite soldiers on rotten wood

The pantropical subfamily Nasutitermitinae has a specialised caste of soldiers, known as nasutes, that have the ability to exude noxious liquids through a horn-like frontal projection that they use for defence. Nasutes have lost their mandibles through the course of evolution and must be fed by workers. A wide variety of monoterpene hydro-carbon solvents have been identified in the liquids that nasutes secrete.

Soldiers of the species *Globitermes sulphureus* commit suicide by autothysis – rupturing a large gland just beneath the surface of their cuticles. The thick, yellow fluid in the gland becomes very sticky on contact with the air, entangling ants or other insects which are trying to invade the nest. Another termite, *Neocapriterme taracua*, also engages in suicidal defence. Workers physically unable to use their mandibles while in a fight form a pouch full of chemicals, then deliberately rupture themselves, releasing toxic chemicals that paralyse and kill their enemies. The soldiers of the neotropical termite family Serritermitidae have a defence strategy which involves front gland autothysis, with the body rupturing between the head and abdomen. When soldiers guarding nest entrances are attacked by intruders, they engage in autothysis, creating a block that denies entry to any attacker.

Workers use several different strategies to deal with their dead, including burying, cannibalism, and avoiding a corpse altogether. To avoid pathogens, termites occasionally engage in necrophoresis, in which a nestmate will carry away a corpse from the colony to dispose of it elsewhere. Which strategy is used depends on the nature of the corpse a worker is dealing with (i.e. the age of the carcass).

Relationship with Other Organisms

A species of fungus is known to mimic termite eggs, successfully avoiding its natural predators. These small brown balls, known as "termite balls", rarely kill the eggs, and in some cases the workers will even tend to them. This fungus mimics these eggs by producing a cellulose-digesting enzyme known as glucosidases. A unique mimicking behaviour exists between various species of *Trichopsenius* beetles and certain termite species within *Reticulitermes*. The beetles share the same cuticle hydrocarbons as the termites and even biosynthesize them. This chemical mimicry allows the beetles to integrate themselves within the termite colonies. The developed appendages on the physogastric abdomen of *Austrospirachtha mimetes* allows the beetle to mimic a termite worker.

Rhizanthella gardneri is the only known orchid in the world pollinated by termites.

Some species of ant are known to capture termites to use as a fresh food source later on, rather than killing them. For example, *Formica nigra* captures termites, and those who try to escape are immediately seized and driven underground. Certain species of ants in the subfamily Ponerinae conduct these raids although other ant species go in alone to steal the eggs or nymphs. Ants such as *Megaponera analis* attack the outside the mounds and Dorylinae ants attack underground. Despite this, some termites and ants can coexist peacefully. Some species of termite, including *Nasutitermes corniger*, form associations with certain ant species to keep away predatory ant species. The earliest known association between *Azteca* ants and *Nasutitermes* termites date back to the Oligocene to Miocene period.

An ant raiding party collecting Pseudocanthotermes militaris termites after a successful raid

54 species of ants are known to inhabit *Nasutitermes* mounds, both occupied and abandoned ones. One reason many ants live in *Nasutitermes* mounds is due to the termites'

frequent occurrence in their geographical range; another is to protect themselves from floods. *Iridomyrmex* also inhabits termite mounds although no evidence for any kind of relationship (other than a predatory one) is known. In rare cases, certain species of termites live inside active ant colonies. Some invertebrate organisms such as beetles, caterpillars, flies and millipedes are termitophiles and dwell inside termite colonies (they are unable to survive independently). As a result, certain beetles and flies have evolved with their hosts. They have developed a gland that secrete a substance that attracts the workers by licking them. Mounds may also provide shelter and warmth to birds, lizards, snakes and scorpions.

Termites are known to carry pollen and regularly visit flowers, so are regarded as potential pollinators for a number of flowering plants. One flower in particular, *Rhizanthella gardneri*, is regularly pollinated by foraging workers, and it is perhaps the only Orchidaceae flower in the world to be pollinated by termites.

Many plants have developed effective defences against termites. However, seedlings are vulnerable to termite attacks and need additional protection, as their defence mechanisms only develop when they have passed the seedling stage. Defence is typically achieved by secreting antifeedant chemicals into the woody cell walls. This reduces the ability of termites to efficiently digest the cellulose. A commercial product, "Blockaid", has been developed in Australia that uses a range of plant extracts to create a paint-on nontoxic termite barrier for buildings. An extract of a species of Australian figwort, *Eremophila*, has been shown to repel termites; tests have shown that termites are strongly repelled by the toxic material to the extent that they will starve rather than consume the food. When kept close to the extract, they become disoriented and eventually die.

Nests

A termite nest can be considered as being composed of two parts, the inanimate and the animate. The animate is all of the termites living inside the colony, and the inanimate part is the structure itself, which is constructed by the termites. Nests can be broadly separated into three main categories: subterranean (completely below ground), epigeal (protruding above the soil surface), and arboreal (built above ground, but always connected to the ground via shelter tubes). Epigeal nests (mounds) protrude from the earth with ground contact and are made out of earth and mud. A nest has many functions such as providing a protected living space and providing shelter against predators. Most termites construct underground colonies rather than multifunctional nests and mounds. Primitive termites of today nest in wooden structures such as logs, stumps and the dead parts of trees, as did termites millions of years ago.

To build their nests, termites primarily use faeces, which have many desirable properties as a construction material. Other building materials include partly digested plant material, used in carton nests (arboreal nests built from faecal elements and wood), and soil, used in subterranean nest and mound construction. Not all nests are visible, as many nests in tropical forests are located underground. Species in the subfamily Apicotermitinae are good examples of subterranean nest builders, as they only dwell inside tunnels. Other termites live in wood, and tunnels are constructed as they feed on the wood. Nests and mounds protect the termites' soft bodies against desiccation, light, pathogens and parasites, as well as providing a fortification against predators. Nests made out of carton are particularly weak, and so the inhabitants use counter-attack strategies against invading predators.

An arboreal termite nest in Mexico

Some species build complex nests called polycalic nests; this habitat is called polycalism. Polycalic species of termites form multiple nests, or calies, connected by subterranean chambers. The termite genera *Apicotermes* and *Trinervitermes* are known to have polycalic species. Polycalic nests appear to be less frequent in mound-building species although polycalic arboreal nests have been observed in a few species of *Nasutitermes*.

Mounds

Nests are considered mounds if they protrude from the earth's surface. A mound provides termites the same protection as a nest but is stronger. Mounds located in areas with torrential and continuous rainfall are at risk of mound erosion due to their clay-rich construction. Those made from carton can provide protection from the rain, and in fact can withstand high precipitation. Certain areas in mounds are used as strong points in case of a breach. For example, *Cubitermes* colonies build narrow tunnels used as strong points, as the diameter of the tunnels is small enough for soldiers to block. A highly protected chamber, known as the "queens cell", houses the queen and king and is used as a last line of defence.

Species in the genus *Macrotermes* arguably build the most complex structures in the insect world, constructing enormous mounds. These mounds are among the largest in the world, reaching a height of 8 to 9 metres (26 to 29 feet), and consist of chimneys, pinnacles and ridges. Another termite species, *Amitermes meridionalis*, can build nests 3 to 4 metres (9 to 13 feet) high and 2.5 metres (8 feet) wide.

Cathedral mounds in the Northern Territory, Australia

The sculptured mounds sometimes have elaborate and distinctive forms, such as those of the compass termite (*Amitermes meridionalis* and *A. laurensis*), which builds tall, wedge-shaped mounds with the long axis oriented approximately north–south, which gives them their common name. This orientation has been experimentally shown to assist thermo-regulation. The north-south orientation causes the internal temperature of a mound to in-crease rapidly during the morning while avoiding overheating from the midday sun. The temperature then remains at a plateau for the rest of the day until the evening.

Mounds of "compass" or "magnetic" termites (*Amitermes*) oriented north-south, thereby avoiding mid-day heat)

Arboreal Nests

Arboreal carton nests of mangrove swamp-dwelling *Nasutitermes* are enriched in lig-nin and depleted in cellulose and xylans. This change is caused by bacterial decay in the gut of the termites: they use their faeces as a carton building material. Arboreal termites nests can account for as much as 2% of above ground carbon storage in Puerto Rican mangrove swamps. These *Nasutitermes* nests are mainly composed of partially biodegraded wood material from the stems and branches of mangrove trees, namely, *Rhizophora mangle* (red mangrove), *Avicennia germinans* (black mangrove) and *La-guncularia racemose* (white mangrove).

Shelter Tubes

Termites construct shelter tubes, also known as earthen tubes or mud tubes, that start from the ground. These shelter tubes can be found on walls and other structures. Constructed by termites during the night, a time of higher humidity, these tubes provide protection to termites from potential predators, especially ants. Shelter tubes also provide high humidity and darkness and allow workers to collect food sources that cannot be accessed in any other way. These passageways are made from soil and faeces and are normally brown in colour. The size of these shelter tubes depends on the amount of food sources that are available. They range from less than 1 cm to several cm in width, but may extend dozens of metres in length.

Nasutiterminae shelter tubes on a tree trunk provide cover for the trail from nest to forest floor.

Relationship with Humans

As Pests

Owing to their wood-eating habits, many termite species can do great damage to unprotected buildings and other wooden structures. Their habit of remaining concealed often results in their presence being undetected until the timbers are severely damaged, leaving a thin layer of a wall that protects them from the environment. Of the 3,106 species known, only 183 species cause damage; 83 species cause significant damage to wooden structures. In North America, nine subterranean species are pests; in Australia, 16 species have an economic impact; in the Indian subcontinent 26 species are considered pests, and in tropical Africa, 24. In Central America and the West Indies, there are 17 pest species. Among the termite genera, *Coptotermes* has the highest number of pest species of any genus, with 28 species known to cause damage. Less than 10% of drywood termites are pests, but they infect wooden structures and furniture in tropical, subtropical and other regions. Dampwood termites only attack lumber material exposed to rainfall or soil.

Termite damage on external structure

Drywood termites thrive in warm climates, and human activities can enable them to invade homes since they can be transported through contaminated goods, containers and ships. Colonies of termites have been seen thriving in warm buildings located in cold regions. Some termites are considered invasive species. *Cryptotermes brevis*, the most widely introduced invasive termite species in the world, has been introduced to all the islands in the West Indies and to Australia.

Termite damage in wooden house stumps

In addition to causing damage to buildings, termites can also damage food crops. Termites may attack trees whose resistance to damage is low but generally ignore fast-growing plants. Most attacks occur at harvest time; crops and trees are attacked during the dry season.

The damage caused by termites costs the southwestern United States approximately $1.5 billion each year in wood structure damage, but the true cost of damage worldwide cannot be determined. Drywood termites are responsible for a large proportion of the damage caused by termites.

To better control the population of termites, various methods have been developed to track termite movements. One early method involved distributing termite bait laced with immunoglobulin G (IgG) marker proteins from rabbits or chickens. Termites collected from the field could be tested for the rabbit-IgG markers using a rabbit-IgG-specific assay. More recently developed, less expensive alternatives include tracking the termites using egg white, cow milk, or soy milk proteins, which can be sprayed on termites in the field. Termites bearing these proteins can be traced using a protein-specific ELISA test.

As Food

Mozambican boys from the Yawo tribe collecting flying termites

These flying alates were collected as they came out of their nests in the ground during the early days of the rainy season.

43 termite species are used as food by humans or are fed to livestock. These insects are particularly important in less developed countries where malnutrition is common, as the protein from termites can help improve the human diet. Termites are consumed in many regions globally, but this practice has only become popular in developed nations in recent years.

Termites are consumed by people in many different cultures around the world. In Africa, the alates are an important factor in the diets of native populations. Tribes have different ways of collecting or cultivating insects; sometimes tribes will collect soldiers from several species. Though harder to acquire, queens are regarded as a delicacy. Termite alates are high in nutrition with adequate levels of fat and protein. They are regarded as pleasant in taste, having a nut-like flavour after they are cooked.

Alates are collected when the rainy season begins. During a nuptial flight, they are typically seen around lights to which they are attracted, and so nets are set up on lamps and captured alates are later collected. The wings are removed through a technique that is similar to winnowing. The best result comes when they are lightly roasted on a hot plate or fried until crisp. Oil is not required as their bodies usually contain sufficient amounts of oil. Termites are typically eaten when livestock is lean and tribal crops have not yet developed or produced any food, or if food stocks from a previous growing season are limited.

In addition to Africa, termites are consumed in local or tribal areas in Asia and North and South America. In Australia, Indigenous Australians are aware that termites are edible but do not consume them even in times of scarcity; there are few explanations as to why. Termite mounds are the main sources of soil consumption (geophagy) in many countries including Kenya, Tanzania, Zambia, Zimbabwe and South Africa. Researchers have suggested that termites are suitable candidates for human consumption and space agriculture, as they are high in protein and can be used to convert inedible waste to consumable products for humans.

In Agriculture

Termites can be major agricultural pests, particularly in East Africa and North Asia, where crop losses can be severe (3–100% in crop loss in Africa). Counterbalancing this

is the greatly improved water infiltration where termite tunnels in the soil allow rain-water to soak in deeply, which helps reduce runoff and consequent soil erosion through bioturbation. In South America, cultivated plants such as eucalyptus, upland rice and sugarcane can be severely damaged by termite infestations, with attacks on leaves, roots and woody tissue. Termites can also attack other plants, including cassava, coffee, cotton, fruit trees, maize, peanuts, soybeans and vegetables. Mounds can disrupt farming activities, making it difficult for farmers to operate farming machinery; however, despite farmers' dislike of the mounds, it is often the case that no net loss of production occurs. Termites can be beneficial to agriculture, such as by boosting crop yields and enriching the soil. Termites and ants can re-colonise untilled land that contains crop stubble, which colonies use for nourishment when they establish their nests. The presence of nests in fields enables larger amounts of rainwater to soak into the ground and increases the amount of nitrogen in the soil, both essential for the growth of crops.

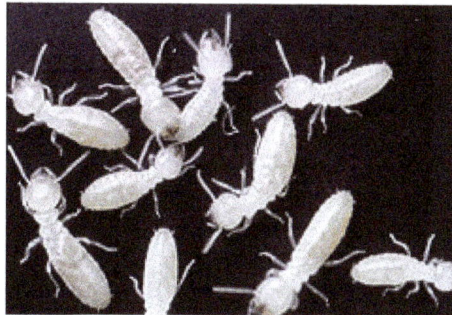

Scientists have developed a more affordable method of tracing the movement of termites using traceable proteins.

In Science and Technology

The termite gut has inspired various research efforts aimed at replacing fossil fuels with cleaner, renewable energy sources. Termites are efficient bioreactors, capable of producing two litres of hydrogen from a single sheet of paper. Approximately 200 species of microbes live inside the termite hindgut, releasing the hydrogen that was trapped inside wood and plants that they digest. Through the action of unidentified enzymes in the termite gut, lignocellulose polymers are broken down into sugars and are transformed into hydrogen. The bacteria within the gut turns the sugar and hydrogen into cellulose acetate, an acetate ester of cellulose on which termites rely for energy. Community DNA sequencing of the microbes in the termite hindgut has been employed to provide a better understanding of the metabolic pathway. Genetic engineering may enable hydrogen to be generated in bioreactors from woody biomass.

The development of autonomous robots capable of constructing intricate structures without human assistance has been inspired by the complex mounds that termites build. These robots work independently and can move by themselves on a tracked grid, capable of climbing and lifting up bricks. Such robots may be useful for future projects on Mars, or for building levees to prevent flooding.

Termites use sophisticated means to control the temperatures of their mounds. As discussed above, the shape and orientation of the mounds of the Australian compass termite stabilises their internal temperatures during the day. As the towers heat up, the solar chimney effect (stack effect) creates an updraft of air within the mound. Wind blowing across the tops of the towers enhances the circulation of air through the mounds, which also include side vents in their construction. The solar chimney effect has been in use for centuries in the Middle East and Near East for passive cooling, as well as in Europe by the Romans. It is only relatively recently, however, that climate responsive construction techniques have become incorporated into modern architecture. Especially in Africa, the stack effect has become a popular means to achieve natural ventilation and passive cooling in modern buildings.

In Culture

The Eastgate Centre is a shopping centre and office block in central Harare, Zimbabwe, whose architect, Mick Pearce, used passive cooling inspired by that used by the local termites. It was the first major building exploiting termite-inspired cooling techniques to attract international attention. Other such buildings include the Learning Resource Center at the Catholic University of Eastern Africa and the Council House 2 building in Melbourne, Australia.

The pink-hued Eastgate Centre

Few zoos hold termites, due to the difficulty in keeping them captive and to the reluctance of authorities to permit potential pests. One of the few that do, the Zoo Basel in Switzerland, has two thriving *Macrotermes bellicosus* populations – resulting in an event very rare in captivity: the mass migrations of young flying termites. This happened in September 2008, when thousands of male termites left their mound each night, died, and covered the floors and water pits of the house holding their exhibit.

African tribes in several countries have termites as totems, and for this reason tribe members are forbidden to eat the reproductive alates. Termites are widely used in traditional popular medicine; they are used as treatments for diseases and other conditions such as asthma, bronchitis, hoarseness, influenza, sinusitis, tonsillitis and whooping cough. In Nigeria, *Macrotermes nigeriensis* is used for spiritual protection and to treat wounds and sick pregnant women. In Southeast Asia, termites are used in ritual practices. In Malaysia, Singapore and Thailand, termite mounds are commonly worshiped among the populace. Abandoned mounds are viewed as structures created by spirits, believing a local guardian dwells within the mound; this is known as Keramat and Datok Kong. In urban areas, local residents construct red-painted shrines over mounds that have been abandoned, where they pray for good health, protection and luck.

Lepidoptera

The Lepidoptera is an order of insects that includes moths and butterflies (both called lepidopterans). 180,000 species of Lepidoptera are described, in 126 families and 46 superfamilies, 10% of the total described species of living organisms. It is one of the most widespread and widely recognizable insect orders in the world, encompassing moths and the three superfamilies of butterflies, skipper butterflies, and moth-butterflies. The Lepidoptera show many variations of the basic body structure that have evolved to gain advantages in lifestyle and distribution. Recent estimates suggest the order may have more species than earlier thought, and is among the four most speciose orders, along with the Hymenoptera, Diptera, and Coleoptera.

Lepidopteran species are characterized by more than three derived features, some of the most apparent being the scales covering their bodies and wings, and a proboscis. The scales are modified, flattened "hairs", and give butterflies and moths their extraordinary variety of colors and patterns. Almost all species have some form of membranous wings, except for a few that have reduced wings or are wingless. Like most other insects, butterflies and moths are holometabolous, meaning they undergo complete metamorphosis. Mating and the laying of eggs are carried out by adults, normally near or on host plants for the larvae. The larvae are commonly called caterpillars, and are completely different from their adult moth or butterfly forms, having a cylindrical body with a well-developed head, mandible mouth parts, three pairs of thoracic legs and from none up to five pairs of prolegs. As they grow, these larvae change in appearance, going through a series of stages called instars. Once fully matured, the larva develops into a pupa, referred to as a chrysalis in the case of butterflies and a cocoon in the case of moths. A few butterflies and many moth species spin a silk case or cocoon prior to pupating, while others do not, instead going underground.

The Lepidoptera have, over millions of years, evolved a wide range of wing patterns and coloration ranging from drab moths akin to the related order Trichoptera, to the brightly colored and complex-patterned butterflies. Accordingly, this is the most recognized and popular of insect orders with many people involved in the observation, study, collection, rearing of, and commerce in these insects. A person who collects or studies this order is referred to as a lepidopterist.

Butterflies and moths play an important role in the natural ecosystem as pollinators and as food in the food chain; conversely, their larvae are considered very problematic to vegetation in agriculture, as their main source of food is often live plant matter. In many species, the female may produce from 200 to 600 eggs, while in others, the number may approach 30,000 eggs in one day. The caterpillars hatching from these eggs can cause damage to large quantities of crops. Many moth and butterfly species are of economic interest by virtue of their role as pollinators, the silk they produce, or as pest species.

Etymology

The origins of the common names "butterfly" and "moth" are varied and often obscure. The English word butterfly is from Old English *buttorfleoge*, with many variations in spelling. Other than that, the origin is unknown, although it could be derived from the pale yellow color of many species' wings suggesting the color of butter. The species of Heterocera are commonly called moths. The origins of the English word moth are more clear, deriving from the Old English *moððe*" (cf. Northumbrian dialect *mohðe*) from Common Germanic (compare Old Norse *motti*, Dutch *mot* and German *Motte* all meaning "moth"). Perhaps its origins are related to Old English *maða* meaning "maggot" or from the root of "midge", which until the 16th century was used mostly to indicate the larva, usually in reference to devouring clothes.

The etymological origins of the word "caterpillar", the larval form of butterflies and moths, are from the early 16th century, from Middle English *catirpel*, *catirpeller*, probably an alteration of Old North French *catepelose*: *cate*, cat (from Latin *cattus*) + *pelose*, hairy.

Distribution and Diversity

The Lepidoptera are among the most successful groups of insects. They are found on all continents, except Antarctica, and inhabit all terrestrial habitats ranging from desert to rainforest, from lowland grasslands to montane plateaus, but almost always associated with higher plants, especially angiosperms (flowering plants). Among the most northern dwelling species of butterflies and moths is the Arctic Apollo (*Parnassius arcticus*), which is found in the Arctic Circle in northeastern Yakutia, at an altitude of 1500 meters above sea level. In the Himalayas, various Apollo species such as *Parnassius epaphus* have been recorded to occur up to an altitude of 6,000 m above sea level.

Some lepidopteran species exhibit symbiotic, phoretic, or parasitic lifestyles, inhabiting the bodies of organisms rather than the environment. Coprophagous pyralid moth species, called sloth moths, such as *Bradipodicola hahneli* and *Cryptoses choloepi*, are unusual in that they are exclusively found inhabiting the fur of sloths, mammals found in Central and South America. Two species of *Tinea* moths have been recorded as feeding on horny tissue and have been bred from the horns of cattle. The larva of *Zenodochium coccivorella* is an internal parasite of the coccid *Kermes* species. Many species have been recorded as breeding in natural materials or refuse such as owl pellets, bat caves, honeycombs or diseased fruit.

As of 2007, there was roughly 174,250 lepidopteran species described, with butterflies and skippers estimated to comprise around 17,950, and moths making up the rest. The vast majority of Lepidoptera are to be found in the tropics, but substantial diversity exists on most continents. North America has over 700 species of butterflies and over 11,000 species of moths, while about 400 species of butterflies and 14,000 species of moths are reported from Australia. The diversity of Lepidoptera in each faunal region has been estimated by John Heppner in 1991 based partly on actual counts from the literature, partly on the card indices in the Natural History Museum (London) and the National Museum of Natural History (Washington), and partly on estimates:

Diversity of Lepidoptera in each faunal region					
	Palearctic	**Nearctic**	**Neotropic**	**Afrotropic**	**Indo-Australian (comprising Indomalayan and Australian regions)**
Estimated number of species	22,465	11,532	44,791	20,491	47,286

External Morphology

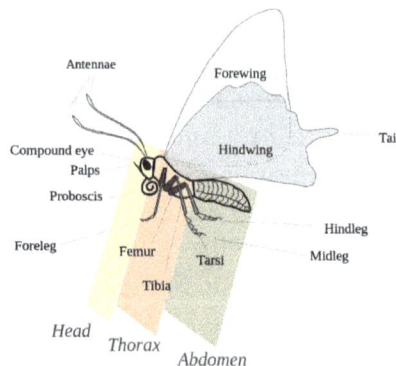

Parts of an adult butterfly

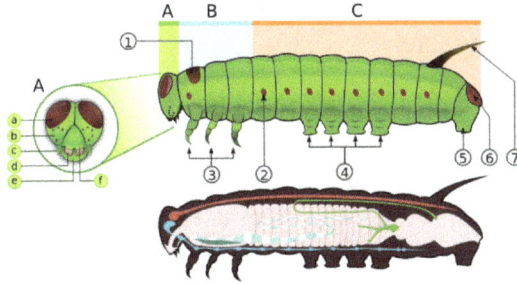

A – head, B – thorax, C – abdomen, 1 – prothoracic shield, 2 – spiracle, 3 – true legs, 4 – midabdominal prolegs, 5 – anal proleg, 6 – anal plate, 7 – tentacle, a – frontal triangle, b – stemmata (ocelli), c – antenna, d – mandible, e – labrum.

Lepidoptera are morphologically distinguished from other orders principally by the presence of scales on the external parts of the body and appendages, especially the wings. Butterflies and moths vary in size from microlepidoptera only a few millimeters long, to conspicuous animals with a wingspan greater than 25 centimetres, such as the Monarch butterfly and Atlas moth. Lepidopterans undergo a four-stage lifecycle: egg; larva or caterpillar; pupa or chrysalis; and imago (plural: imagines) / adult and show many variations of the basic body structure, which have evolved to gain advantages in lifestyle and distribution.

Head

The face of a caterpillar with the mouthparts showing

The head is where many sensing organs and the mouth parts are found. Like the adult, the larva also has a toughened, or sclerotized head capsule. Here, two compound eyes, and *chaetosema*, raised spots or clusters of sensory bristles unique to Lepidoptera, occur, though many taxa have lost one or both of these spots. The antennae have a wide variation in form among species and even between different sexes. The antennae of butterflies are usually filiform and shaped like clubs, those of the skippers are hooked, while those of moths have flagellar segments variously enlarged or branched. Some moths have enlarged antennae or ones that are tapered and hooked at the ends.

The maxillary galeae are modified and form an elongated proboscis. The proboscis consists of one to five segments, usually kept coiled up under the head by small muscles when it is not being used to suck up nectar from flowers or other liquids. Some basal moths still have mandibles, or separate moving jaws, like their ancestors, and these form the family Micropterigidae.

The larvae, called caterpillars, have a toughened head capsule. Caterpillars lack the proboscis and have separate chewing mouthparts. These mouthparts, called mandibles, are used to chew up the plant matter that the larvae eat. The lower jaw, or labium, is weak, but may carry a spinneret, an organ used to create silk. The head is made of large lateral lobes, each having an ellipse of up to six simple eyes.

Thorax

The thorax is made of three fused segments, the prothorax, mesothorax, and metathorax, each with a pair of legs. The first segment contains the first pair of legs. In some males of the butterfly family Nymphalidae, the fore legs are greatly reduced and are not used for walking or perching. The three pairs of legs are covered with scales. Lepidoptera also have olfactory organs on their feet, which aid the butterfly in "tasting" or "smelling" out its food. In the larval form there are 3 pairs of true legs, with up to 11 pairs of abdominal legs (usually eight) and hooklets, called apical crochets.

The two pairs of wings are found on the middle and third segments, or mesothorax and metathorax, respectively. In the more recent genera, the wings of the second segment are much more pronounced, although some more primitive forms have similarly sized wings of both segments. The wings are covered in scales arranged like shingles, which form an extraordinary variety of colors and patterns. The mesothorax is designed to have more powerful muscles to propel the moth or butterfly through the air, with the wing of this segment (fore wing) having a stronger vein structure. The largest superfamily, the Noctuidae, has their wings modified to act as tympanal or hearing organs.

The caterpillar has an elongated, soft body that may have hair-like or other projections, three pairs of true legs, with none to 11 pairs of abdominal legs (usually eight) and hooklets, called apical crochets. The thorax usually has a pair of legs on each segment. The thorax is also lined with many spiracles on both the mesothorax and metathorax, except for a few aquatic species, which instead have a form of gills.

Abdomen

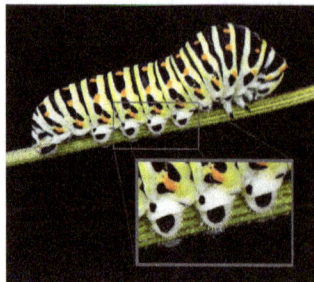

Caterpillar prolegs on Papilio machaon

The abdomen, which is less sclerotized than the thorax, consists of 10 segments with membranes in between, allowing for articulated movement. The sternum, on the first segment, is small in some families and is completely absent in others. The last two or three segments form the external parts of the species' sex organs. The genitalia of Lepidoptera are highly varied and are often the only means of differentiating between species. Male genitals include a valva, which is usually large, as it is used to grasp the female during mating. Female genitalia include three distinct sections.

The females of basal moths have only one sex organ, which is used for copulation and as an ovipositor, or egg-laying organ. About 98% of moth species have a separate organ for mating, and an external duct that carries the sperm from the male.

The abdomen of the caterpillar has four pairs of prolegs, normally located on the third to sixth segments of the abdomen, and a separate pair of prolegs by the anus, which have a pair of tiny hooks called crotchets. These aid in gripping and walking, especially in species that lack many prolegs (e. g. larvae of Geometridae). In some basal moths, these prolegs may be on every segment of the body, while prolegs may be lost completely in other groups, which are more adapted to boring and living in sand (e. g., Prodoxidae and Nepticulidae, respectively).

Scales

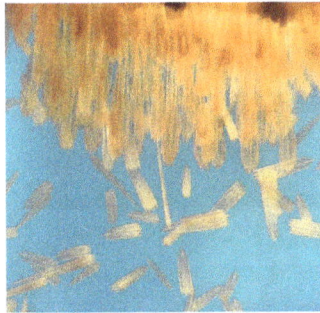

Wing scales form the color and pattern on wings. The scales shown here are lamellar. The pedicel can be seen attached to a few loose scales.

The wings, head, and parts of the thorax and abdomen of Lepidoptera are covered with minute scales, a feature from which the order derives its name. Most scales are lamellar, or blade-like and attached with a pedicel, while other forms may be hair-like or specialized as secondary sexual characteristics.

The lumen or surface of the lamella has a complex structure. It gives color either by colored pigments it contains, or through structural coloration with mechanisms that include photonic crystals and diffraction gratings.

Scales function in insulation, thermoregulation, producing pheromones (in males only), and aiding gliding flight, but the most important is the large diversity of vivid or indistinct patterns they provide, which help the organism protect itself by camouflage or mimicry, and which act as signals to other animals including rivals and potential mates.

Electron Microscopy Images of Scales

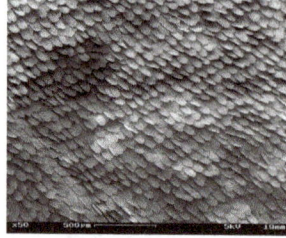

A patch of wing (×50)

Scales close up (×200)

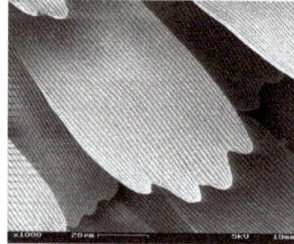

A single scale (×1000)

Microstructure of a scale (×5000)

Internal Morphology

Reproductive System

In the reproductive system of butterflies and moths, the male genitalia are complex and unclear. In females the three types of genitalia are based on the relating taxa: 'monotrysian', 'exoporian', and 'ditrysian'. In the monotrysian type is an opening on the fused segments of the sterna 9 and 10, which act as insemination and oviposition. In the exoporian type (in Hepaloidae and Mnesarchaeoidea) are two separate places for insemination and oviposition, both occurring on the same sterna as the monotrysian type, i.e. 9 and 10. The ditrysian groups have an internal duct that carries sperm, with separate openings for copulation and egg-laying. In most species, the genitalia are flanked by two soft lobes, although they may be specialized and sclerotized in some species for ovipositing in area such as crevices and inside plant tissue. Hormones and the glands that produce them run the development of butterflies and moths as they go through their lifecycles, called the endocrine system. The first insect hormone prothoracicotropic hormone(PTTH) operates the species lifecycle and diapause. This hormone is produced by corpora allata and corpora cardiaca, where it is also stored. Some glands are specialized to perform certain task such as producing silk or producing saliva in the palpi. While the corpora cardiaca produce PTTH, the corpora allata also produces juvenile hormones, and the prothorocic glands produce moulting hormones.

Digestive System

In the digestive system, the anterior region of the foregut has been modified to form a pharyngeal sucking pump as they need it for the food they eat, which are for the most part liquids. An esophagus follows and leads to the posterior of the pharynx and in some species forms a form of crop. The midgut is short and straight, with the hindgut being longer and coiled. Ancestors of lepidopteran species, stemming from Hymenoptera, had midgut ceca, although this is lost in current butterflies and moths. Instead, all the digestive enzymes, other than initial digestion, are immobilized at the surface of the midgut cells. In larvae, long-necked and stalked goblet cells are found in the anterior and posterior midgut regions, respectively. In insects, the goblet cells excrete positive potassium ions, which are absorbed from leaves ingested by the larvae. Most butterflies and moths display the usual digestive cycle, but species with different diets require adaptations to meet these new demands.

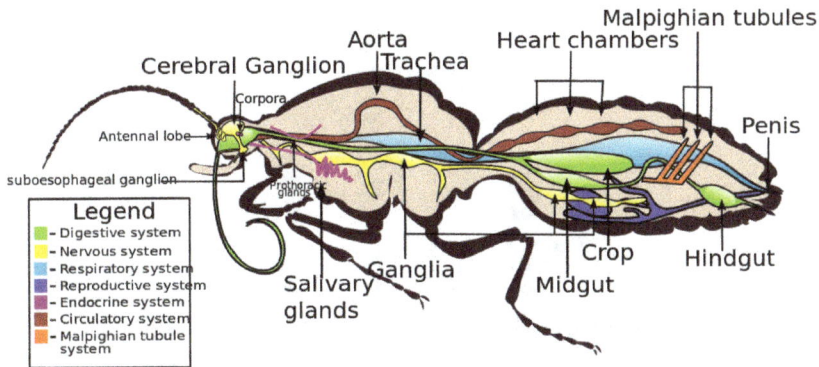

Internal morphology of adult male in the family Nymphalidae, showing most of the major organ systems, with characteristic reduced forelegs of that family: The corpora include the corpus allatum and the corpus cardiaca.

Circulatory System

In the circulatory system, hemolymph, or insect blood, is used to circulate heat in a form of thermoregulation, where muscles contraction produces heat, which is transferred to the rest of the body when conditions are unfavorable. In lepidopteran species, hemolymph is circulated through the veins in the wings by some form of pulsating organ, either by the heart or by the intake of air into the trachea.

Respiratory System

Air is taken in through spiracles along the sides of the abdomen and thorax supplying the trachea with oxygen as it goes through the lepidopteran's respiratory system. Three different tracheaes supply and diffuse oxygen throughout the species' bodies. The dorsal tracheae supply oxygen to the dorsal musculature and vessels, while the ventral tracheae supply the ventral musculature and nerve cord, and the visceral tracheae supply the guts, fat bodies, and gonads.

Polymorphism

Sexually dimorphic bagworm moths (Thyridopteryx ephemeraeformis) mating: The female is flightless.

The *Heliconius* butterflies from the tropics of the Western Hemisphere are the classical model for Müllerian mimicry.

Polymorphism is the appearance of forms or "morphs", which differ in color and number of attributes within a single species. In Lepidoptera, polymorphism can be seen not only between individuals in a population, but also between the sexes as sexual dimorphism, between geographically separated populations in geographical polymorphism, and between generations flying at different seasons of the year (seasonal polymorphism or polyphenism). In some species, the polymorphism is limited to one sex, typically the female. This often includes the phenomenon of mimicry when mimetic morphs fly alongside nonmimetic morphs in a population of a particular species. Polymorphism occurs both at specific level with heritable variation in the overall morphological design of individuals, as well as in certain specific morphological or physiological traits within a species.

Environmental polymorphism, in which traits are not inherited, is often termed as polyphenism, which in Lepidoptera is commonly seen in the form of seasonal morphs, especially in the butterfly families of Nymphalidae and Pieridae. An Old World pierid butterfly, the common grass yellow (*Eurema hecabe*) has a darker summer adult morph, triggered by a long day exceeding 13 hours in duration, while the shorter diurnal period of 12 hours or less induces a paler morph in the postmonsoon period. Polyphenism also occurs in caterpillars, an example being the peppered moth, *Biston betularia*.

Geographical isolation causes a divergence of a species into different morphs. A good example is the Indian white admiral *Limenitis procris*, which has five forms, each geographically separated from the other by large mountain ranges. An even more dramatic showcase of geographical polymorphism is the Apollo butterfly (*Parnassius apollo*). Because the Apollos live in small local populations, thus having no contact with each other, coupled with their strong stenotopic nature and weak migration ability, interbreeding between populations of one species practically does not occur; by this, they form over 600 different morphs, with the size of spots on the wings of which varies greatly.

Seasonal Diphenism in the Common Grass Yellow, Eurema Hecabe

Dry-season form

Wet-season form

Sexual dimorphism is the occurrence of differences between males and females in a species. In Lepidoptera, it is widespread and almost completely set by genetic determination. Sexual dimorphism is present in all families of the Papilionoidea and more prominent in the Lycaenidae, Pieridae, and certain taxa of the Nymphalidae. Apart from color variation, which may differ from slight to completely different color-pattern combinations, secondary sexual characteristics may also be present. Different genotypes maintained by natural selection may also be expressed at the same time. Polymorphic and/or mimetic females occur in the case of some taxa in the Papilionidae primarily to obtain a level of protection not available to the male of their species. The most distinct case of sexual dimorphism is that of adult females of many Psychidae species which have only vestigial wings, legs, and mouthparts as compared to the adult males that are strong fliers with well-developed wings and feathery antennae.

Reproduction and Development

Mating pair of Laothoe populi (poplar hawk-moth) showing two different color variants

Species of Lepidoptera undergo holometabolism or "complete metamorphosis". Their lifecycle normally consists of an egg, a larva, a pupa, and an imago or adult. The larvae are commonly called caterpillars, and the pupae of moths encapsulated in silk are called cocoons, while the uncovered pupae of butterflies are called chrysalides.

Lepidopterans in Diapause

Unless the species reproduces year-round, a butterfly or moth may enter diapause, a state of dormancy that allows the insect to survive unfavorable environmental conditions.

Mating

Males usually start eclosion (emergence) earlier than females and peak in numbers before females. Both of the sexes are sexually mature by the time of eclosion. Butterflies and moths normally do not associate with each other, except for migrating species, staying relatively asocial. Mating begins with an adult (female or male) attracting a mate, normally using visual stimuli, especially in diurnal species like most butterflies. However, the females of most nocturnal species, including almost all moth species, use pheromones to attract males, sometimes from long distances. Some species engage in a form of acoustic courtship, or attract mates using sound or vibration such as the polka-dot wasp moth, *Syntomeida epilais*.

Adaptations include undergoing one seasonal generation, two or even more, called voltinism (Univoltism, bivoltism, and multivism, respectively). Most lepidopterans in temperate climates are univoltine, while in tropical climates most have two seasonal broods. Some others may take advantage of any opportunity they can get, and mate continuously throughout the year. These seasonal adaptations are controlled by hormones, and these delays in reproduction are called diapause. Many lepidopteran species, after mating and laying their eggs, die shortly afterwards, having only lived for a few days after eclosion. Others may still be active for several weeks and then overwinter and become sexually active again when the weather becomes more favorable, or diapause. The sperm of the male that mated most recently with the female is most likely to have fertilized the eggs, but the sperm from a prior mating may still prevail.

Lifecycle

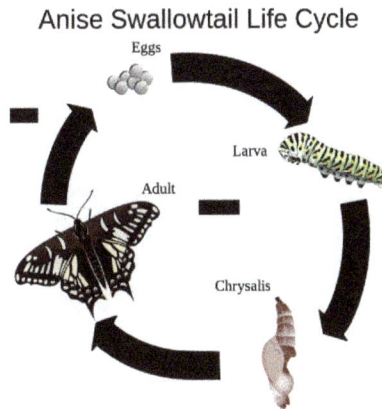

The four stages of the lifecycle of an anise swallowtail

Eggs

Lepidoptera usually reproduce sexually and are oviparous (egg-laying), though some species exhibit live birth in a process called ovoviviparity. A variety of differences in egg-laying and the number of eggs laid occur. Some species simply drop their eggs in flight (these species normally have polyphagous larvae, meaning they eat a variety of plants e. g., hepialids and some nymphalids) while most lay their eggs near or on the host plant on which the larvae feed. The number of eggs laid may vary from only a few to several thousand. The females of both butterflies and moths select the host plant instinctively, and primarily, by chemical cues.

The eggs are derived from materials ingested as a larvae and in some species, from the spermataphores received from males during mating. An egg can only be 1/1000 the mass of the female, yet she may lay up to her own mass in eggs. Females lay smaller eggs as they age. Larger Females lay larger eggs. The egg is covered by a hard-ridged protective outer layer of shell, called the chorion. It is lined with a thin coating of wax, which prevents the egg from drying out. Each egg contains a number of micropyles, or tiny funnel-shaped openings at one end, the purpose of which is to allow sperm to enter and fertilize the egg. Butterfly and moth eggs vary greatly in size between species, but they are all either spherical or ovate.

The egg stage lasts a few weeks in most butterflies, but eggs laid prior to winter, especially in temperate regions, go through diapause, and hatching may be delayed until spring. Other butterflies may lay their eggs in the spring and have them hatch in the summer. These butterflies are usually temperate species (e. g. *Nymphalis antiopa*).

Larvae

Larval form typically lives and feeds on plants

The larvae or caterpillars are the first stage in the lifecycle after hatching. Caterpillars, are "characteristic polypod larvae with cylindrical bodies, short thoracic legs, and abdominal prolegs (pseudopods)". They have a toughened (sclerotised) head capsule with an adfrontal suture formed by medial fusion of the sclerites, mandibles (mouthparts) for chewing, and a soft tubular, segmented body, that may have hair-like or other projections, three pairs of true legs, and additional prolegs (up to five pairs). The body consists of thirteen segments, of which three are thoracic and ten are abdominal. Most larvae are herbivores, but a few are carnivores (some eat ants or other caterpillars) and detritivores.

Different herbivorous species have adapted to feed on every part of the plant and are normally considered pests to their host plants; some species have been found to lay their eggs on the fruit and other species lay their eggs on clothing or fur (e. g., *Tineola bisselliella*, the common clothes moth). Some species are carnivorous and others are even parasitic. Some lycaenid species such as *Maculinea rebeli* are social parasites of *Myrmica* ants nests. A species of Geometridae from Hawaii has carnivorous larvae that catch and eat flies. Some pyralid caterpillars are aquatic.

The larvae develop rapidly with several generations in a year; however, some species may take up to 3 years to develop, and exceptional examples like *Gynaephora groen-landica* take as long as seven years. The larval stage is where the feeding and growing stages occur, and the larvae periodically undergo hormone-induced ecdysis, developing further with each instar, until they undergo the final larval-pupal molt. The lepidopteran pupa, known as a chrysalis in the case of butterflies, has functional mandibles with appendages fused or glued to the body in most species, while the pupal mandibles are not functional in others.

The larvae of both butterflies and moths exhibit mimicry to deter potential predators. Some caterpillars have the ability to inflate parts of their heads to appear snake-like. Many have false eye-spots to enhance this effect. Some caterpillars have special structures called osmeteria (family Papilionidae), which are exposed to produce smelly chemicals used in defense. Host plants often have toxic substances in them and caterpillars are able to sequester these substances and retain them into the adult stage. This helps make them unpalatable to birds and other predators. Such unpalatability is advertised using bright red, orange, black, or white warning colors. The toxic chemicals in plants are often evolved specifically to prevent them from being eaten by insects. Insects, in turn, develop countermeasures or make use of these toxins for their own survival. This "arms race" has led to the coevolution of insects and their host plants.

Wing Development

No form of wing is externally visible on the larva, but when larvae are dissected, developing wings can be seen as disks, which can be found on the second and third thoracic segments, in place of the spiracles that are apparent on abdominal segments. Wing disks develop in association with a trachea that runs along the base of the wing, and are surrounded by a thin peripodial membrane, which is linked to the outer epidermis of the larva by a tiny duct. Wing disks are very small until the last larval instar, when they increase dramatically in size, are invaded by branching tracheae from the wing base that precede the formation of the wing veins, and begin to develop patterns associated with several landmarks of the wing.

Near pupation, the wings are forced outside the epidermis under pressure from the hemolymph, and although they are initially quite flexible and fragile, by the time the pupa breaks free of the larval cuticle, they have adhered tightly to the outer cuticle of the pupa (in obtect pupae). Within hours, the wings form a cuticle so hard and well-joined to the body that pupae can be picked up and handled without damage to the wings.

Pupa

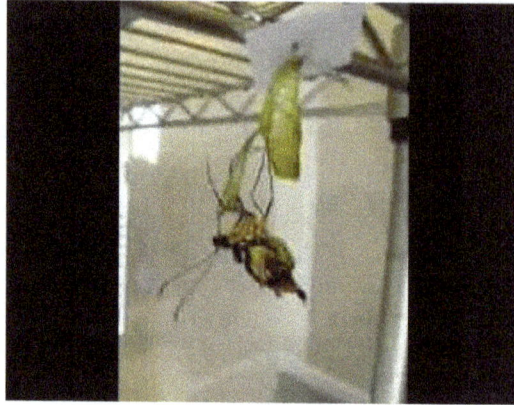

Eclosion of Papilio dardanus

After about five to seven instars, or molts, certain hormones, like PTTH, stimulate the production of ecdysone, which initiates insect molting. Then, the larva puparium, a sclerotized or hardened cuticle of the last larval instar, develops into the pupa. Depending on the species, the pupa may be covered in silk and attached to many different types of debris, or may not be covered at all. The pupa stays attached to the leaf by silk spun by the caterpillar before it spins the silk for the full pupa. Features of the imago are externally recognizable in the pupa. All the appendages on the adult head and thorax are found cased inside the cuticle (antennae, mouthparts, etc.), with the wings wrapped around, adjacent to the antennae.

While encased, some of the lower segments are not fused, and are able to move using small muscles found in between the membrane. Moving may help the pupa, for example, escape the sun, which would otherwise kill it. The pupa of the Mexican jumping bean moth (*Cydia deshaisiana*) does this. The larvae cut a trapdoor in the bean (species of *Sebastiania*) and use the bean as a shelter. With a sudden rise in temperature, the pupa inside twitches and jerks, pulling on the threads inside. Wiggling may also help to deter parasitoid wasps from laying eggs on the pupa. Other species of moths are able to make clicks to deter predators.

The length of time before the pupa ecloses (emerges) varies greatly. The monarch butterfly may stay in its chrysalis for two weeks, while other species may need to stay for more than 10 months in diapause. The adult emerges from the pupa either by using abdominal hooks or from projections located on the head. The mandibles found in the most primitive moth families are used to escape from their cocoon (e. g., Micropterigoidea).

Adult

Most lepidopteran species do not live long after eclosion, only needing a few days to find a mate and then lay their eggs. Others may remain active for a longer period (from one to several weeks), or go through diapause and overwintering as monarch butterflies do, or waiting out environmental stress. Some adult species of Micro-lepidoptera go through a stage where no reproductive-related activity occurs, lasting through summer and winter, followed by mating and oviposition in the early spring.

While most butterflies and moths are terrestrial, many species of Pyralidae are truly aquatic with all stages except the adult occurring in water. Many species from other families such as Arctiidae, Nepticulidae, Cosmopterygidae, Tortricidae, Olethreutidae, Noctuidae, Cossidae, and Sphingidae are aquatic or semiaquatic.

Behavior

Flight

Flight is an important aspect of the lives of butterflies and moths, and is used for evading predators, searching for food, and finding mates in a timely manner, as lepidopteran species do not live long after eclosion. It is the main form of locomotion in most species. In Lepidoptera, the fore wings and hind wings are mechanically coupled and flap in synchrony. Flight is anteromotoric, or being driven primarily by action of the fore wings. Although lepidopteran species reportedly can still fly when their hind wings are cut off, it reduces their linear flight and turning capabilities.

Lepidopteran species have to be warm, about 77 to 79 °F (25 to 26 °C), to fly. They depend on their body temperature being sufficiently high and since they cannot regulate it themselves, this is dependent on their environment. Butterflies living in cooler climates may use their wings to warm their bodies. They will bask in the sun, spreading out their wings so that they get maximum exposure to the sunlight. In hotter climates butterflies can easily overheat, so they are usually active only during the cooler parts of the day, early morning, late afternoon or early evening. During the heat of the day, they rest in the shade. Some larger thick-bodied moths (e.g. Sphingidae) can generate their own heat to a limited degree by vibrating their wings. The heat generated by the flight muscles warms the thorax while the temperature of the abdomen is unimportant for flight. To avoid overheating, some moths rely on hairy scales, internal air sacs, and other structures to separate the thorax and abdomen and keep the abdomen cooler.

Some species of butterflies can reach fast speeds, such as the southern dart, which can go as fast as 48.4 km/h. Sphingids are some of the fastest flying insects, some are capable of flying at over 50 km/h (30 mi/h), having a wingspan of 35–150 mm. In some species, sometimes a gliding component to their flight exists. Flight occurs either as hovering, or as forward or backward motion. In butterfly and moth species, such as hawk moths, hovering is important as they need to maintain a certain stability over flowers when feeding on the nectar.

Navigation

Timelapse of flying moths, attracted to the floodlights

Navigation is important to lepidoptera species, especially for those that migrate. Butterflies, which have more species that migrate, have been shown to navigate using time-compensated sun compasses. They can see polarized light, so can orient even in cloudy conditions. The polarized light in the region close to the ultraviolet spectrum is suggested to be particularly important. Most migratory butterflies are those that live in semiarid areas where breeding seasons are short. The life histories of their host plants also influence the strategies of the butterflies. Other theories include the use of landscapes. Lepidoptera may use coastal lines, mountains, and even roads to orient themselves. Above sea, the flight direction is much more accurate if the coast is still visible.

Many studies have also shown that moths navigate. One study showed that many moths may use the Earth's magnetic field to navigate, as a study of the moth heart and dart suggests. Another study, of the migratory behavior of the silver Y, showed, even at high altitudes, the species can correct its course with changing winds, and prefers flying with favourable winds, suggesting a great sense of direction. *Aphrissa statira* in Panama loses its navigational capacity when exposed to a magnetic field, suggesting it uses the Earth's magnetic field.

Moths exhibit a tendency to circle artificial lights repeatedly. This suggests they use a technique of celestial navigation called transverse orientation. By maintaining a constant angular relationship to a bright celestial light, such as the Moon, they can fly in a straight line. Celestial objects are so far away, even after traveling great distances, the change in angle between the moth and the light source is negligible; further, the moon will always be in the upper part of the visual field or on the horizon. When a moth encounters a much closer artificial light and uses it for navigation, the angle changes noticeably after only a short distance, in addition to being often below the horizon. The moth instinctively attempts to correct by turning toward the light, causing airborne moths to come plummeting downwards, and at close range, which results in a spiral flight path that gets closer and closer to the light source. Other explanations have been suggested, such as the idea that moths may be impaired with a visual distortion called a Mach band by Henry Hsiao in 1972. He stated that they fly towards the darkest part of the sky in pursuit of safety, thus are inclined to circle ambient objects in the Mach band region.

Migration

Monarch butterflies, seen in a cluster in Santa Cruz, California, where the western population migrates for the winter

Lepidopteran migration is typically seasonal, as the insects moving to escape dry seasons or other disadvantageous conditions. Most lepidopterans that migrate are butterflies, and the distance travelled varies. Some butterflies that migrate include the mourning cloak, painted lady, American lady, red admiral, and the common buckeye. The most well-known migrations are those of the eastern population of the monarch butterfly from Mexico to northern United States and southern Canada, a distance of about 4,000–4,800 km (2,500–3,000 mi). Other well-known migratory species include the painted lady and several of the danaine butterflies. Spectacular and large-scale migrations associated with the monsoons are seen in peninsular India. Migrations have been studied in more recent times using wing tags and stable hydrogen isotopes.

Moths also undertake migrations, an example being the uraniids. *Urania fulgens* undergoes population explosions and massive migrations that may be not surpassed by any other insect in the Neotropics. In Costa Rica and Panama, the first population movements may begin in July and early August and depending on the year, may be very massive, continuing unabated for as long as five months.

Communication

Pheromones are commonly involved in mating rituals among species, especially moths, but they are also an important aspect of other forms of communication. Usually, the pheromones are produced by either the male or the female and detected by members of the opposite sex with their antennae. In many species, a gland between the eighth and ninth segments under the abdomen in the female produces the pheromones. Communication can also occur through stridulation, or producing sounds by rubbing various parts of the body together.

Moths are known to engage in acoustic forms of communication, most often as courtship, attracting mates using sound or vibration. Like most other insects, moths pick up these sounds using tympanic membranes in their abdomens. An example is that of the polka-dot wasp moth (*Syntomeida epilais*), which produces sounds with a frequency above that normally detectable by humans (about 20 kHz). These sounds also function as tactile communication, or communication through touch, as they stridulate, or vibrate a substrate like leaves and stems.

Group of Melitaea athalia near Warka, Poland

Most moths lack bright colors, as many species use coloration as camouflage, but butterflies engage in visual communication. Female cabbage butterflies, for example, use ultraviolet light to communicate, with scales colored in this range on the dorsal wing surface. When they fly, each down stroke of the wing creates a brief flash of ultraviolet light which the males apparently recognize as the flight signature of a potential mate. These flashes from the wings may attract several males that engage in aerial courtship displays.

Ecology

Moths and butterflies are important in the natural ecosystem. They are integral participants in the food chain; having co-evolved with flowering plants and predators, lepidopteran species have formed a network of trophic relationships between autotrophs and heterotrophs, which are included in the stages of Lepidoptera larvae, pupae, and adults. Larvae and pupae are links in the diets of birds and parasitic entomophagous insects. The adults are included in food webs in a much broader range of consumers (including birds, small mammals, reptiles, etc.).

Defense and Predation

Lepidopteran species are soft bodied, fragile, and almost defenseless, while the immature stages move slowly or are immobile, hence all stages are exposed to predation. Adult butterflies and moths are preyed upon by birds, lizards, amphibians, dragonflies, and spiders. Caterpillars and pupae fall prey not only to birds, but also to invertebrate predators and small mammals, as well as fungi and bacteria. Parasitoid and parasitic wasps and flies may lay eggs in the caterpillar, which eventually kill it as they hatch inside its body and eat its tissues. Insect-eating birds are probably the largest predators. Lepidoptera, especially the immature stages, are an ecologically important food to many insectivorous birds, such as the great tit in Europe.

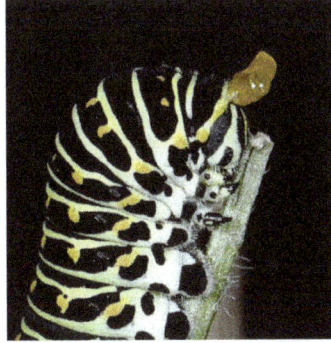

Papilio machaon caterpillar showing the osmeterium, which emits unpleasant smells to ward off predators

An "evolutionary arms race" can be seen between predator and prey species. The Lepidoptera have developed a number of strategies for defense and protection, including evolution of morphological characters and changes in ecological lifestyles and behaviors. These include aposematism, mimicry, camouflage, and development of threat patterns and displays. Only a few birds, such as the nightjars, hunt nocturnal lepidopterans. Their main predators are bats. Again, an "evolutionary race" exists, which has led to numerous evolutionary adaptations of moths to escape from their main predators, such as the ability to hear ultrasonic sounds, or even to emit sounds in some cases. Lepidopteran eggs are also preyed upon. Some caterpillars, such as the zebra swallowtail butterfly larvae, are cannibalistic.

Some species of Lepidoptera are poisonous to predators, such as the monarch butterfly in the Americas, *Atrophaneura* species (roses, windmills, etc.) in Asia, as well as *Papilio antimachus*, and the birdwings, the largest butterflies in Africa and Asia, respectively. They obtain their toxicity by sequestering the chemicals from the plants they eat into their own tissues. Some Lepidoptera manufacture their own toxins. Predators that eat poisonous butterflies and moths may become sick and vomit violently, learning not to eat those species. A predator which has previously eaten a poisonous lepidopteran may avoid other species with similar markings in the future, thus saving many other species, as well. Toxic butterflies and larvae tend to develop bright colors and striking patterns as an indicator to predators about their toxicity. This phenomenon is known as aposematism. Some caterpillars, especially members of Papilionidae, contain an osmeterium, a Y-shaped protrusible gland found in the prothoracic segment of the larvae. When threatened, the caterpillar emits unpleasant smells from the organ to ward off the predators.

Camouflage is also an important defense strategy, which involves the use of coloration or shape to blend into the surrounding environment. Some lepidopteran species blend with their surroundings, making them difficult to spot by predators. Caterpillars can exhibit shades of green that match its host plant. Others look like inedible objects, such as twigs or leaves. For instance, the mourning cloak fades into the backdrop of trees when it folds its wings back. The larvae of some species, such as the common Mormon (*Papilio polytes*) and the western tiger swallowtail look like bird droppings. For example, adult

Sesiidae species (also known as clearwing moths) have a general appearance sufficiently similar to a wasp or hornet to make it likely the moths gain a reduction in predation by Batesian mimicry. Eyespots are a type of automimicry used by some butterflies and moths. In butterflies, the spots are composed of concentric rings of scales in different colors. The proposed role of the eyespots is to deflect attention of predators. Their resemblance to eyes provokes the predator's instinct to attack these wing patterns.

Batesian and Müllerian mimicry complexes are commonly found in Lepidoptera. Genetic polymorphism and natural selection give rise to otherwise edible species (the mimic) gaining a survival advantage by resembling inedible species (the model). Such a mimicry complex is referred to as Batesian and is most commonly known in the example between the limenitidine viceroy butterfly in relation to the inedible danaine monarch. The viceroy is, in fact, more toxic than the monarch and this resemblance should be considered as a case of Müllerian mimicry. In Müllerian mimicry, inedible species, usually within a taxonomic order, find it advantageous to resemble each other so as to reduce the sampling rate by predators that need to learn about the insects' inedibility. Taxa from the toxic genus *Heliconius* form one of the most well-known Müllerian complexes. The adults of the various species now resemble each other so well, the species cannot be distinguished without close morphological observation and, in some cases, dissection or genetic analysis.

Moths evidently are able to hear the range emitted by bat]s, which in effect causes flying moths to make evasive maneuvers because bats are a main predator of moths. Ultrasonic frequencies trigger a reflex action in the noctuid moth that cause it to drop a few inches in its flight to evade attack. Tiger moths in a defense emit clicks within the same range of the bats, which interfere with the bats and foil their attempts to echolocate it.

Pollination

Most species of Lepidoptera engage in some form of entomophily (more specifically psychophily and phalaenophily for butterflies and moths, respectively), or the pollination of flowers. Most adult butterflies and moths feed on the nectar inside flowers, using their probosces to reach the nectar hidden at the base of the petals. In the process, the adults brush against the flowers' stamens, on which the reproductive pollen is made and stored. The pollen is transferred on appendages on the adults, which fly to the next flower to feed and unwittingly deposit the pollen on the stigma of the next flower, where the pollen germinates and fertilizes the seeds.

Flowers pollinated by butterflies tend to be large and flamboyant, pink or lavender in color, frequently having a landing area, and usually scented, as butterflies are typically day-flying. Since butterflies do not digest pollen (except for heliconid species,) more nectar is offered than pollen. The flowers have simple nectar guides, with the nectaries usually hidden in narrow tubes or spurs, reached by the long "tongue" of the butterflies. Butterflies such as *Thymelicus flavus* have been observed to engage in flower constancy, which means they are more likely to transfer pollen to other conspecific plants. This can be beneficial for the plants being pollinated, as flower constancy prevents the loss of pollen during different flights and the pollinators from clogging stigmas with pollen of other flower species.

A day-flying hummingbird hawk-moth drinking nectar from a species of *Dianthus*

Among the more important moth pollinator groups are the hawk moths of the family Sphingidae. Their behavior is similar to hummingbirds, i.e., using rapid wing beats to hover in front of flowers. Most hawk moths are nocturnal or crepuscular, so moth-pollinated flowers (e.g., *Silene latifolia*) tend to be white, night-opening, large, and showy with tubular corollae and a strong, sweet scent produced in the evening, night, or early morning. A lot of nectar is produced to fuel the high metabolic rates needed to power their flight. Other moths (e.g., noctuids, geometrids, pyralids) fly slowly and settle on the flower. They do not require as much nectar as the fast-flying hawk moths, and the flowers tend to be small (though they may be aggregated in heads).

Mutualism

Tobacco hornworm caterpillar Manduca sexta parasitized by Braconidae wasp larvae

Mutualism is a form of biological interaction wherein each individual involved benefits in some way. An example of a mutualistic relationship would be that shared by yucca moths (Tegeculidae) and their host, yucca flowers (Liliaceae). Female yucca moths enter the host flowers, collect the pollen into a ball using specialized maxillary palps, then move to the apex of the pistil, where pollen is deposited on the stigma, and lay eggs into the base of the pistil where seeds will develop. The larvae develop in the fruit pod and feed on a portion of the seeds. Thus, both insect and plant benefit, forming a highly mutualistic relationship. Another form of mutualism occurs between some larvae of butterflies and certain species of ants (e. g. Lycaenidae). The larvae communicate with the ants using vibrations transmitted through a substrate, such as the wood of a tree or stems, as well as using chemical signals. The ants provide some degree of protection to these larvae and they in turn gather honeydew secretions.

Parasitism

Parasitoid larvae exits from the fox moth caterpillar

Only 41 species of parasitoid lepidopterans are known (1-Pyralidae; 40-Epipyropidae) The larvae of the greater and lesser wax moths feed on the honeycomb inside bee nests and may become pests; they are also found in bumblebee and wasp nests, albeit to a lesser extent. In northern Europe, the wax moth is regarded as the most serious parasitoid of the bumblebee, and is found only in bumblebee nests. In some areas in southern England, as much as 80% of nests can be destroyed. Other parasitic larvae are known to prey upon cicadas and leaf hoppers.

In reverse, moths and butterflies may be subject to parasitic wasps and flies, which may lay eggs on the caterpillars, which hatch and feed inside its body, resulting in death. Although, in a form of parasitism called idiobiont, the adult paralyzes the host, so as not to kill it but for it to live as long as possible, for the parasitic larvae to benefit the most. In another form of parasitism, koinobiont, the species live off their hosts while inside (endoparasitic). These parasites live inside the host caterpillar throughout its lifecycle, or may affect it later on as an adult. In other orders, koinobionts include flies, a majority of coleopteran, and many hymenopteran parasitoids. Some species may be subject to a variety of parasites, such as the gypsy moth (*Lymantaria dispar*), which is attacked by a series of 13 species, in six different taxa throughout its lifecycle.

In response to a parsitoid egg or larva in the caterpillar's body, the plasmatocytes, or simply the host's cells can form a multilayered capsule that eventually causes the endoparasite to asphyxiate. The process, called encapsulation, is one of the caterpillar's only means of defense against parasitoids.

Parasitism in Gypsy Moths

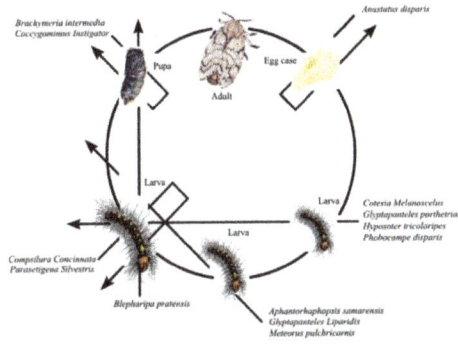

The different parasitoids affecting the gypsy moth (Lymantaria dispar): The stage they affect and eventually kill and its duration are denoted by arrows.

Other Biological Interactions

A few species of Lepidoptera are secondary consumers, or predators. These species typically prey upon the eggs of other insects, aphids, scale insects, or ant larvae. Some caterpillars are cannibals, and others prey on caterpillars of other species (e. g. Hawaiian *Eupithecia*). Those of the 15 species in *Eupithecia* that mirror inchworms, are the only known species of butterflies and moths that are ambush predators. Four species are known to eat snails. For example, the Hawaiian caterpillar, (*H. molluscivora*), uses silk traps, in a manner similar to that of spiders, to capture certain species of snails (typically Tornatellides).

Larvae of some species of moths in the Tineidae, Gelechioidea, and Noctuidae (family/ superfamily/families, respectively), besides others, feed on detritus, or dead organic material, such as fallen leaves and fruit, fungi, and animal products, and turn it into humus. Well-known species include the cloth moths (*Tineola bisselliella, T. pellionella, and T. tapetzella*), which feed on detritus containing keratin, including hair, feathers, cobwebs, bird nests (particularly of domestic pigeons, *Columba livia domestica*) and fruits or vegetables. These species are important to ecosystems as they remove substances that would otherwise take a long time to decompose.

In 2015 it was reported that wasp bracovirus DNA was present in Lepidoptera such as Monarch butterflies, silkworms and moths. These were described in some newspaper articles as examples of a naturally occurring genetically engineered insects.

Evolution and Systematics

History of Study

Linnaeus in *Systema Naturae* (1758) recognized three divisions of the Lepidoptera: *Papilio, Sphinx* and *Phalaena*, with seven subgroups in *Phalaena*. These persist today as 9 of the superfamilies of Lepidoptera. Other works on classification followed including those by Michael Denis & Ignaz Schiffermüller (1775), Johan Christian Fabricius (1775) and Pierre André Latreille (1796). Jacob Hübner described many genera, and the Lepidopteran genera were catalogued by Ferdinand Ochsenheimer and Georg Friedrich Treitschke in a series of volumes on the Lepidopteran fauna of Europe published between 1807 and 1835. Gottlieb August Wilhelm Herrich-Schäffer (several volumes, 1843–1856), and Edward Meyrick (1895) based their classifications primarily on wing venation. Sir George Francis Hampson worked on the 'Microlepidoptera' during this period and Philipp Christoph Zeller published *The Natural History of the Tineinae* also on Microlepidoptera (1855).

Lepidoptera collection in Cherni Osam Natural Sciences Museum, Troyan, Bulgaria

Among the first entomologists to study fossil insects and their evolution was Samuel Hubbard Scudder (1837–1911), who worked on butterflies. He published a study of the Florissant deposits of Colorado, including the exceptionally preserved *Prodryas persephone*. Andreas V. Martynov (1879–1938) recognized the close relationship between Lepidoptera and Trichoptera in his studies on phylogeny.

Major contributions in the 20th century included the creation of the monotrysia and ditrysia (based on female genital structure) by Borner in 1925 and 1939. Willi Hennig (1913–1976) developed the cladistic methodology and applied it to insect phylogeny. Niels P. Kristensen, E. S. Nielsen and D. R. Davis studied the relationships among monotrysian families and Kristensen worked more generally on insect phylogeny and higher Lepidoptera too. While it is often found that DNA-based phylogenies differ from those based on morphology, this has not been the case for the Lepidoptera; DNA phylogenies correspond to a large extent to morphology-based phylogenies.

Many attempts have been made to group the superfamilies of the Lepidoptera into natural groups, most of which fail because one of the two groups is not monophyletic: Microlepidotera and Macrolepidoptera, Heterocera and Rhopalocera, Jugatae and Frenatae, Monotrysia and Ditrysia.

Fossil Record

The fossil record for Lepidoptera is lacking in comparison to other winged species, and tending not to be as common as some other insects in the habitats that are most conducive to fossilization, such as lakes and ponds, and their juvenile stage has only the head capsule as a hard part that might be preserved. The location and abundance of the most common moth species are indicative that mass migrations of moths occurred over the Palaeogene North Sea, which is why there is a serious lack of moth fossils. Yet there are fossils, some preserved in amber and some in very fine sediments. Leaf mines are also seen in fossil leaves, although the interpretation of them is tricky.

1887 engraving of Prodryas persephone, a fossil Lepidopteran from the Eocene.

Putative fossil stem group representatives of Amphiesmenoptera (the clade comprising Trichoptera and Lepidoptera) are known from the Triassic. The earliest known fossil lepidopteran is *Archaeolepis mane* from the Jurassic, about 190 million years ago in Dorset, UK. The fossil belongs to a small primitive moth-like species, and its wings are showing scales with parallel grooves under a scanning electron microscope and a characteristic wing venation pattern shared with Trichoptera (Caddisflies). Only two more sets of Jurassic lepidopteran fossils have been found, as well as 13 sets from the Cretaceous, which all belong to primitive moth-like families. Many more fossils are found from the Tertiary, and particularly the Eocene Baltic amber. The oldest genuine butterflies of the superfamily Papilionoidea have been found in the Paleocene MoClay or Fur Formation of Denmark. The best preserved fossil lepidopteran is the Eocene *Prodryas persephone* from the Florissant Fossil Beds.

Phylogeny

Lepidoptera and Trichoptera (caddisflies) are more closely related to each other than to any other insect order, sharing many similarities that are lacking in others; for example the females of both orders are heterogametic, meaning they have two different sex chromosomes, whereas in most species the males are heterogametic and the females have two identical sex chromosomes. The adults in both orders display a particular wing venation pattern on their forewings. The larvae of both orders have mouth structures and gland with which they make and manipulate silk. Willi Hennig grouped the two sister orders into the Amphiesmenoptera superorder. This group probably evolved in the Jurassic, having split from the now extinct order Necrotaulidae. Lepidoptera descend from a diurnal moth-like common ancestor that either fed on dead or living plants.

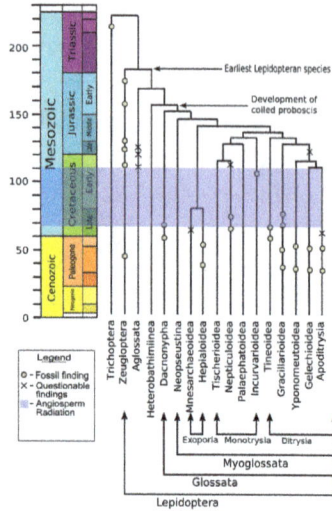

Phylogenetic hypothesis of major lepidopteran lineages superimposed on the geologic time scale. Angiosperm radiation spans 130 to 95 million years ago from the earliest angiosperms, to angiosperm domination of vegetation.

Micropterigidae, Agathiphagidae and Heterobathmiidae are the oldest and most basal lineages of Lepidoptera. The adults of these families do not have the curled tongue or proboscis, that are found in most members order, but instead have chewing mandibles adapted for a special diet. Micropterigidae larvae feed on leaves, fungi, or liverworts (much like the Trichoptera). Adult Micropterigidae chew the pollen or spores of ferns. In the Agathiphagidae, larvae live inside kauri pines and feed on seeds. In Heterobathmiidae the larvae feed on the leaves of *Nothofagus*, the southern beech tree. These families also have mandibles in the pupal stage, which help the pupa emerge from the seed or cocoon after metamorphosis.

The Eriocraniidae have a short coiled proboscis in the adult stage, and though they retain their pupal mandibles with which they escaped the cocoon, their mandibles are non-functional thereafter. Most of these non-ditrysian families, are primarily leaf miners in the larval stage. In addition to the proboscis, there is a change in the scales among these basal lineages, with later lineages showing more complex perforated scales.

With the evolution of the Ditrysia in the mid-Cretaceous, there was a major reproductive change. The Ditrysia, which comprise 98% of the Lepidoptera, have two separate openings for reproduction in the females (as well as a third opening for excretion), one for mating, and one for laying eggs. The two are linked internally by a seminal duct. (In more basal lineages there is one cloaca, or later, two openings and an external sperm canal.) Of the early lineages of Ditrysia, Gracillarioidea and Gelechioidea are mostly leaf miners, but more recent lineages feed externally. In the Tineoidea, most species feed on plant and animal detritus and fungi, and build shelters in the larval stage.

The Yponomeutoidea is the first group to have significant numbers of species whose larvae feed on herbaceous plants, as opposed to woody plants. They evolved about the time that flowering plants underwent an expansive adaptive radiation in the mid-Cretaceous, and the Gelechioidea that evolved at this time also have great diversity. Whether the processes involved coevolution or sequential evolution, the diversity of the Lepidoptera and the angiosperms increased together.

In the so-called "Macrolepidoptera", which constitutes about 60% of lepidopteran species, there was a general increase in size, better flying ability (via changes in wing shape and linkage of the forewings and hindwings), reduction in the adult mandibles, and a change in the arrangement of the crochets (hooks) on the larval prolegs, perhaps to improve the grip on the host plant. Many also have tympanal organs, that allow them to hear. These organs evolved eight times, at least, because they occur on different body parts and have structural differences. The main lineages in the Macrolepidoptera are the Noctuoidea, Bombycoidea, Lasiocampidae, Mimallonoidea, Geometroidea and Rhopalocera. Bombycoidea plus Lasiocampidae plus Mimallonoidea may be a monophyletic group. The Rhopalocera, comprising the Papilionoidea (butterflies), Hesperioidea (skippers), and the Hedyloidea (moth-butterflies), are the most recently evolved. There is quite a good fossil record for this group, with the oldest skipper dating from 56 million years ago.

Taxonomy

Taxonomy is the classification of species in selected taxa, the process of naming being called nomenclature. There are over 120 families in lepidoptera, in 45 to 48 superfamilies. Lepidoptera have always been, historically, classified in five suborders, one of which is of primitive moths that never lost the morphological features of its ancestors. The rest of the moths and butterflies make up ninety-eight percent of the other taxa, making Ditrysia. More recently, findings of new taxa, larvae and pupa have aided in detailing the relationships of primitive taxa, phylogenetic analysis showing the primitive lineages to be paraphyletic compared to the rest of Lepidoptera lineages. Recently lepidopterists have abandoned clades like suborders, and those between orders and superfamilies.

- is a clade with Micropterigoidea being its only superfamily, containing the single family Micropterigidae. Species of Micropterigoidea are practically living fossils, being one of the most primitive lepidopteran groups, still retaining chewing mouthparts (mandibles) in adults, unlike other clades of butterflies and moths. About 120 species are known worldwide, with more than half the species in the genus *Micropteryx* in the Paleartic region. There are only 2 known in North America (*Epimartyria*), with many more being found Asia and the southwest Pacific, particularly New Zealand with about 50 species.

- contains a majority of the species, with the most obvious difference is non-functioning mandibles, and elongated maxillary galeae or the proboscis. The basal clades still retaining some of the ancestral features of the wings such as similarly shaped fore- and hindwings with relatively complete venation. Glossata also contains the division Ditrysia, which contains 98% of all described species in Lepidoptera.

- it is the second most primitive lineage of lepidoptera; being first described in 1952 by Lionel Jack Dumbleton. Agathiphagidae and Heterobathmiidae are the only families in Aglossata. Agathiphagidae only contains about 2 species in its genus Agathiphaga. *Agathiphaga queenslandensis* and *Agathiphaga vitiensis*, being found along the north-eastern coast of Queensland, Australia, and in Fiji to Vanuatu and the Solomon Islands, respectively.

- Heterobathmiina was first described by Kristensen and Nielsen in 1979. There are about 10 species, which are day-flying, metallic moths, confined to southern South America, the adults eat the pollen of *Nothofagus* or Southern Beech and the larvae mine the leaves.

Relationship to People

Culture

Death's-head Hawkmoth (*Acherontia lachesis*), an old bleached specimen still showing the classical skull pattern on the thorax

Artistic depictions of butterflies have been used in many cultures including as early as 3500 years ago, in Egyptian hieroglyphs. Today, butterflies are widely used in various objects of art and jewelry: mounted in frames, embedded in resin, displayed in bottles, laminated in paper, and in some mixed media artworks and furnishings. Butterflies have also inspired the "butterfly fairy" as an art and fictional character.

In many cultures the soul of a dead person is associated with the butterfly, for example in Ancient Greece, where the word for butterfly also means *soul* and *breath*. In Latin, as in Ancient Greece, the word for "butterfly" papilio was associated with the soul of the dead. The skull-like marking on the thorax of the Death's-head Hawkmoth has helped these moths, particularly *A. atropos*, earn a negative reputation, such as associations with the supernatural and evil. The moth has been prominently featured in art and movies such as *Un Chien Andalou* (by Buñuel and Dalí) and *The Silence of the Lambs*, and in the artwork of the Japanese metal band Sigh's album *Hail Horror Hail*. According to *Kwaidan: Stories and Studies of Strange Things*, by Lafcadio Hearn, a butterfly was seen in Japan as the personification of a person's soul;

whether they be living, dying, or already dead. One Japanese superstition says that if a butterfly enters your guestroom and perches behind the bamboo screen, the person whom you most love is coming to see you. However, large numbers of butterflies are viewed as bad omens. When Taira no Masakado was secretly preparing for his famous revolt, there appeared in Kyoto so vast a swarm of butterflies that the people were frightened—thinking the apparition to be a portent of coming evil.

In the ancient Mesoamerican city of Teotihuacan, the brilliantly colored image of the butterfly was carved into many temples, buildings, jewelry, and emblazoned on incense burners in particular. The butterfly was sometimes depicted with the maw of a jaguar and some species were considered to be the reincarnations of the souls of dead warriors. The close association of butterflies to fire and warfare persisted through to the Aztec civilization and evidence of similar jaguar-butterfly images has been found among the Zapotec, and Maya civilizations.

Caterpillar hatchling of the Grey Dagger (Acronicta psi) eating leaves from a tree

Pests

The larvae of many Lepidopteran species are major pests in agriculture. Some of the major pests include Tortricidae, Noctuidae, and Pyralidae. The larvae of the Noctuidae genus *Spodoptera* (armyworms), *Helicoverpa* (corn earworm), or *Pieris brassicae* can cause extensive damage to certain crops. *Helicoverpa zea* larvae (cotton bollworms or tomato fruitworms) are polyphagous, meaning they eat a variety of crops, including tomatoes and cotton.

Butterflies and moths are one of the largest taxa to solely feed and be dependent on living plants, in terms of the number of species, and they are in many ecosystems, making up the largest biomass to do so. In many species, the female may produce anywhere from 200 to 600 eggs, while in some others it may go as high as 30,000 eggs in one day. This can create many problems for agriculture, where many caterpillars can affect acres of vegetation. Some reports estimate that there have been over 80,000 caterpillars of several different taxa feeding on a single oak tree. In some cases, phytophagous larvae can lead to the destruction of entire trees in relatively short periods of time.

Ecological ways of removing pest lepidoptera species are becoming more economically viable, as research has shown ways like introducing parasitic wasp and flies. For example, *Sarcophaga aldrichi*, a fly which deposited larvae feed upon the pupae of the Forest Tent Caterpillar Moth. Pesticides can affect other species other than the species they are targeted to eliminate, damaging the natural ecosystem. Another good biological pest control method is the use of pheromone traps. A pheromone trap is a type of insect trap that uses pheromones to lure insects. Sex pheromones and aggregating pheromones are the most common types used. A pheromone-impregnated lure is encased in a conventional trap such as a Delta trap, water-pan trap, or funnel trap.

Species of moths that are detritivores would naturally eat detritus containing keratin, such as hairs or feathers. Well known species are cloth moths (*T. bisselliella*, *T. pellionella*, and *T. tapetzella*), feeding on foodstuffs that people find economically important, such as cotton, linen, silk and wool fabrics as well as furs; furthermore they have been found on shed feathers and hair, bran, semolina and flour (possibly preferring wheat flour), biscuits, casein, and insect specimens in museums.

Beneficial Insects

Even though most butterflies and moths affect the economy negatively, some species are a valuable economic resource. The most prominent example is that of the Domesticated silkworm moth (*Bombyx mori*), the larvae of which make their cocoons out of silk, which can be spun into cloth. Silk is and has been an important economic resource throughout history. The species *Bombyx mori* has been domesticated to the point where it is completely dependent on mankind for survival. A number of wild moths such as *Bombyx mandarina*, and *Antheraea* species, besides others, provide commercially important silks.

The preference of the larvae of most Lepidopteran species to feed on a single species or limited range of plants is used as a mechanism for biological control of weeds in place of herbicides. The pyralid cactus moth was introduced from Argentina to Australia, where it successfully suppressed millions of acres of Prickly pear cactus. Another species of the Pyralidae, called the alligator weed stem borer (*Arcola malloi*), was used to control the aquatic plant known as alligator weed (*Alternanthera philoxeroides*) in conjunction with the alligator weed flea beetle; in this case, the two insects work in synergy and the weed rarely recovers.

Breeding butterflies and moths, or butterfly gardening/rearing, has become an ecologically viable process of introducing species into the ecosystem to benefit it. Butterfly ranching in Papua New Guinea permits nationals of that country to "farm" economically valuable insect species for the collectors market in an ecologically sustainable manner.

Food

Lepidoptera feature prominently in entomophagy as food items on almost every continent. While in most cases, adults, larvae or pupae are eaten as staples by indigenous people, beondegi or silkworm pupae are eaten as a snack in Korean cuisine while Maguey worm is considered a delicacy in Mexico. In the Carnia region of Italy, children catch and eat ingluvies of the toxic *Zygaena* moths in early summer. The ingluvies, despite having a very low cyanogenic content, serve as a convenient, supplementary source of sugar to the children who can include this resource as a seasonal delicacy at minimum risk.

Beondegi, silkworm pupae steamed or boiled and seasoned for taste, for sale by a street vendor in South Korea

Health

Some larvae of both moths and butterflies have a form of hair that has been known to be a cause of human health problems. Caterpillar hairs sometimes have toxins in them and species from approximately 12 families of moths or butterflies worldwide can inflict serious human injuries (Urticarial dermatitis and atopic asthma to osteochondritis, consumption coagulopathy, renal failure, and intracerebral hemorrhage). Skin rashes are the most common, but there have been fatalities. *Lonomia* is a frequent cause of envenomation in humans in Brazil, with 354 cases reported between 1989 and 2005. Lethality ranging up to 20% with death caused most often by intracranial hemorrhage.

These hairs have also been known to cause kerato-conjunctivitis. The sharp barbs on the end of caterpillar hairs can get lodged in soft tissues and mucous membranes such as the eyes. Once they enter such tissues, they can be difficult to extract, often exacerbating the problem as they migrate across the membrane. This becomes a particular problem in an indoor setting. The hairs easily enter buildings through ventilation systems and accumulate in indoor environments because of their small size, which makes it difficult for them to be vented out. This accumulation increases the risk of human contact in indoor environments.

Bee

Bees are flying insects closely related to wasps and ants, known for their role in pollination and, in the case of the best-known bee species, the European honey bee, for producing honey and beeswax. Bees are a monophyletic lineage within the superfamily Apoidea, presently considered as a clade Anthophila. There are nearly 20,000 known species of bees in seven to nine recognized families, though many are undescribed and the actual number is probably higher. They are found on every continent except Antarctica, in every habitat on the planet that contains insect-pollinated flowering plants.

Some species including honey bees, bumblebees, and stingless bees live socially in colonies. Bees are adapted for feeding on nectar and pollen, the former primarily as an energy source and the latter primarily for protein and other nutrients. Most pollen is used as food for larvae. Bee pollination is important both ecologically and commercially; the decline in wild bees has increased the value of pollination by commercially managed hives of honey bees.

Bees range in size from tiny stingless bee species whose workers are less than 2 millimetres (0.08 in) long, to *Megachile pluto*, the largest species of leafcutter bee, whose females can attain a length of 39 millimetres (1.54 in). The most common bees in the Northern Hemisphere are the Halictidae, or sweat bees, but they are small and often mistaken for wasps or flies. Vertebrate predators of bees include birds such as bee-eaters; insect predators include beewolves and dragonflies.

Human beekeeping or apiculture has been practised for millennia, since at least the times of Ancient Egypt and Ancient Greece. Apart from honey and pollination, honey bees produce beeswax, royal jelly and propolis. Bees have appeared in mythology and folklore, again since ancient times, and they feature in works of literature as varied as Virgil's *Georgics*, Beatrix Potter's *The Tale of Mrs Tittlemouse*, and W. B. Yeats's poem *The Lake Isle of Innisfree*. Bee larvae are included in the Javanese dish *botok tawon*, where they are eaten steamed with shredded coconut.

Evolution

The ancestors of bees were wasps in the family Crabronidae, which were predators of other insects. The switch from insect prey to pollen may have resulted from the consumption of prey insects which were flower visitors and were partially covered with pollen when they were fed to the wasp larvae. This same evolutionary scenario may have occurred within the vespoid wasps, where the pollen wasps evolved from predatory ancestors. Until recently, the oldest non-compression bee fossil had been found in New Jersey amber, *Cretotrigona prisca* of Cretaceous age, a corbiculate bee. A bee fossil from the early Cretaceous (~100 mya), *Melittosphex burmensis*, is considered *"an extinct lineage of pollen-collecting Apoidea sister to the modern bees"*. Derived features of its morphology (apomorphies) place it clearly within the bees, but it retains two unmodified ancestral traits (plesiomorphies) of the legs (two mid-tibial spurs, and a slender hind basitarsus), showing its transitional status. By the Eocene (~45 mya) there was already considerable diversity among eusocial bee lineages.

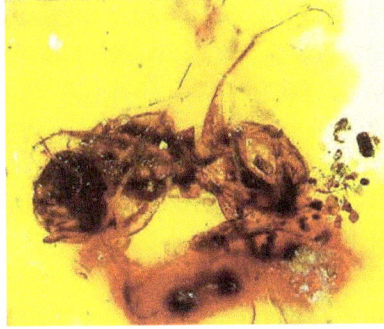

Melittosphex burmensis, a fossil bee preserved in amber from the Early Cretaceous of Myanmar

The highly eusocial corbiculate Apidae appeared roughly 87 Mya, and the Allodapini (within the Apidae) around 53 Mya. The Colletidae appear as fossils only from the late Oligocene (~25 Mya) to early Miocene. The Melittidae are known from *Palaeomacropis eocenicus* in the Early Eocene. The Megachilidae are known from trace fossils (characteristic leaf cuttings) from the Middle Eocene. The Andrenidae are known from the Eocene-Oligocene boundary, around 34 Mya, of the Florissant shale. The Halictidae first appear in the Early Eocene with species found in amber. The Stenotritidae are known from fossil brood cells of Pleistocene age.

Coevolution

Long-tongued bees and long-tubed flowers coevolved, like this Amegilla cingulata (Apidae) on Acanthus ilicifolius.

The earliest animal-pollinated flowers were shallow, cup-shaped blooms pollinated by insects such as beetles, so the syndrome of insect pollination was well established before the first appearance of bees. The novelty is that bees are specialized as pollination agents, with behavioral and physical modifications that specifically enhance pollination, and are the most efficient pollinating insects. In a process of coevolution, flowers developed floral rewards such as nectar and longer tubes, and bees developed longer tongues to extract the nectar. Bees also developed structures known as scopal hairs and pollen baskets to collect and carry pollen. The location and type differ among and between groups of bees. Most bees have scopal hairs located on their hind legs or on the underside of their abdomens, some bees in the family Apidae possess pollen baskets on their hind legs while very few species lack these entirely and instead collect pollen in their crops. This drove the adaptive radiation of the angiosperms, and, in turn, the bees themselves. Bees have not only coevolved with flowers but it is believed that some bees have coevolved with mites. Some bees provide tufts of hairs called acarinaria that appear to provide lodgings for mites; in return, it is believed that the mites eat fungi that attack pollen, so the relationship in this case may be mutualistc.

Phylogeny

External

This cladogram is based on Debevic et al. 2012, which used molecular phylogeny to demonstrate that the bees arose from deep within the Crabronidae, which is therefore paraphyletic. The Heterogynaidae is also broken up.

Internal

This cladogram of the bee families is based on Hedtke et al., 2013, which places the former families Dasypodaidae and Meganomiidae as subfamilies inside the Melittidae. English names, where available, are given in parentheses.

Description

The lapping mouthparts of a honeybee, showing labium and maxillae

It is usually easy to recognise that a particular insect is a bee. They differ from closely related groups such as wasps by having branched or plume-like setae (bristles), combs on the forelimbs for cleaning their antennae, small anatomical differences in the limb structure and the venation of the hind wings, and in females, by having the seventh dorsal abdominal plate divided into two half-plates.

Behaviourally, one of the most obvious characteristics of bees is that they collect pollen to provide provisions for their young, and have the necessary adaptations to do this. However, certain wasp species such as pollen wasps have similar behaviours, and a few species of bee scavenge from carcases to feed their offspring. The world's largest species of bee is thought to be the Indonesian resin bee *Megachile pluto*, whose females can attain a length of 39 millimetres (1.54 in). The smallest species may be dwarf stingless bees in the tribe Meliponini whose workers are less than 2 millimetres (0.08 in) in length.

Head-on view of a carpenter bee, showing antennae, three ocelli, compound eyes, sensory bristles and mouthparts

A bee has a pair of large compound eyes which cover much of the surface of the head. Between and above these are three small simple eyes (ocelli) which provide information for the bee on light intensity. The antennae usually have thirteen segments in males and twelve in females and are geniculate, having an elbow joint part way along. They house large numbers of sense organs that can detect touch (mechanoreceptors), smell and taste, and small, hairlike mechanoreceptors that can detect air movement so as to "hear" sounds. The mouthparts are adapted for both chewing and sucking by having both a pair of mandibles and a long proboscis for sucking up nectar.

The thorax has three segments, each with a pair of robust legs, and a pair of membranous wings on the hind two segments. The front legs of corbiculate bees bear combs for cleaning the antennae, and in many species the hind legs bear pollen baskets, flattened sections with incurving hairs to secure the collected pollen. The wings are synchronised in flight and the somewhat smaller hind wings connect to the forewings by a row of hooks along their margin which connect to a groove in the forewing. The abdomen has nine segments, the hindermost three being modified into the sting.

Sociality

Haplodiploid Breeding System

Willing to die for their sisters: worker honey bees killed defending their hive against wasps, along with a dead wasp. Such eusocial behaviour is favoured by the haplodiploid sex determination system of bees.

According to inclusive fitness theory, organisms can gain fitness not just through increasing their own reproductive output, but also that of close relatives. In evolutionary terms, individuals should help relatives when *Cost < Relatedness * Benefit*. The requirements for eusociality are more easily fulfilled by haplodiploid species such as bees because of their unusual relatedness structure. In haplodiploid species, females develop from fertilized eggs and males from unfertilized eggs. Because a male is haploid (has only one copy of each gene), his daughters (which are diploid, with two copies of each gene) share 100% of his genes and 50% of their mother's. Therefore, they share 75% of their genes with each other. This mechanism of sex determination gives rise to what W. D. Hamilton termed "supersisters", more closely related to their sisters than they would be to their own offspring. Workers often do not reproduce, but they can pass on more of their genes by helping to raise their sisters (as queens) than they would by having their own offspring (each of which would only have 50% of their genes). This unusual situation has been proposed as an explanation of the multiple independent evolutions of eusociality (arising at least nine separate times) within the Hymenoptera. However, some eusocial species such as termites are not haplodiploid. Conversely, many bees are haplodiploid yet are not eusocial, and among eusocial species many queens mate with multiple males, creating half-sisters that share only 25% of their genes. Haplodiploidy is thus neither necessary nor sufficient for eusociality. But, monogamy (queens mating singly) is the ancestral state for all eusocial species so far investigated, so it is likely that haplodiploidy contributed to the evolution of eusociality in bees.

Eusociality

Bees may be solitary or may live in various types of communities. The most advanced of these are eusocial colonies found among the honey bees, bumblebees, and stingless bees; these are characterised by having cooperative brood care and a division of labour into reproductive and non-reproductive adults. Sociality, of several different types, is believed to have evolved separately many times within the bees. In some species, groups of cohabiting females may be sisters, and if there is a division of labour within the group, they are considered semisocial. The group is called eusocial if, in addition, the group consists of a mother (the queen) and her daughters (workers), with male drones at certain stages. When the castes are purely behavioural alternatives, the system is considered primitively eusocial as in many paper wasps; when the castes are morphologically discrete, the system is considered highly eusocial.

There are many more species of primitively eusocial than highly eusocial bees, but they have rarely been studied. Most are in the family Halictidae, or "sweat bees". Colonies are typically small, with a dozen or fewer workers, on average. Queens and workers differ only in size, if at all. Most species have a single season colony cycle, even in the tropics, and only mated females hibernate. A few species have long active seasons and attain colony sizes in the hundreds. The orchid bees include some primitively eusocial species with similar biology. Some allodapine bees are primitively eusocial colonies, with progressive provisioning: a larva's food is supplied gradually as it develops, as is the case in honey bees and some bumblebees.

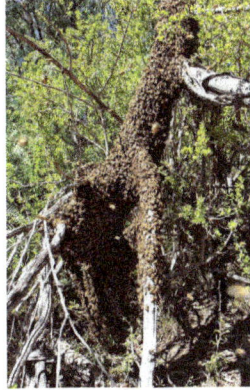

A honey bee swarm

Bumblebees are eusocial, like the eusocial Vespidae such as hornets. The queen initiates a nest on her own. Bumblebee colonies typically have from 50 to 200 bees at peak population, which occurs in mid to late summer. Nest architecture is simple, limited by the size of the pre-existing nest cavity, and colonies rarely last more than a year. In 2011, the International Union for Conservation of Nature set up the Bumblebee Specialist Group to review the threat status of all bumblebee species world-wide using the IUCN Red List criteria.

A bumblebee carrying pollen in its pollen baskets (corbiculae)

Stingless bees are highly eusocial. They practise mass provisioning, with complex nest architecture and perennial colonies.

The true honey bees (genus *Apis*) are highly eusocial, and are among the best known of all insects. There are 29 subspecies of *Apis mellifera*, native to Europe, the Middle East, and Africa. Africanized bees are a hybrid strain of *A. mellifera* that escaped from experiments involving crossing European and African subspecies; they are unusually defensive.

Solitary and Communal Bees

Most other bees, including familiar insects such as carpenter bees, leafcutter bees and mason bees are solitary in the sense that every female is fertile, and typically inhabits a nest she constructs herself. There is no division of labor so these nests lack queens and *worker* bees for these species. Solitary bees typically produce neither honey nor beeswax.

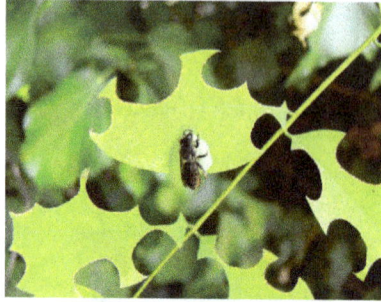

A leafcutting bee, Megachile rotundata cutting circles from acacia leaves

A solitary bee, Anthidium florentinum (family Megachilidae), visiting Lantana

Solitary bees are important pollinators; they gather pollen to provision their nests with food for their brood. Often it is mixed with nectar to form a paste-like consistency. Some solitary bees have advanced types of pollen-carrying structures on their bodies. A very few species of solitary bees are being cultured for commercial pollination. Most of these species belong to a distinct set of genera which are commonly known by their nesting behavior or preferences, namely: carpenter bees, sweat bees, mason bees, polyester bees, squash bees, dwarf carpenter bees, leafcutter bees, alkali bees and digger bees.

The mason bee Osmia cornifrons nests in a hole in dead wood. Bee "hotels" are often sold for this purpose.

Most solitary bees nest in the ground in a variety of soil textures and conditions while others create nests in hollow reeds or twigs, holes in wood. The female typically creates a compartment (a "cell") with an egg and some provisions for the resulting larva, then seals it off. A nest may consist of numerous cells. When the nest is in wood, usually the last (those closer to the entrance) contain eggs that will become males. The adult does not provide care for the brood once the egg is laid, and usually dies after making one or more nests. The males typically emerge first and are ready for mating when the females emerge. Solitary bees are either stingless or very unlikely to sting (only in self-defense, if ever).

While solitary females each make individual nests, some species. such as the European mason bee *Hoplitis anthocopoides*, and the Dawson's Burrowing bee, *Amegilla dawsoni*, are gregarious, preferring to make nests near others of the same species, and giving the appearance of being social. Large groups of solitary bee nests are called *aggregations*, to distinguish them from colonies. In some species, multiple females share a common nest, but each makes and provisions her own cells independently. This type of group is called "communal" and is not uncommon. The primary advantage appears to be that a nest entrance is easier to defend from predators and parasites when there are multiple females using that same entrance on a regular basis.

Biology

Nest of the common carder bumblebee. The wax canopy has been removed to show winged workers and pupae in irregularly placed wax cells.

Life Cycle

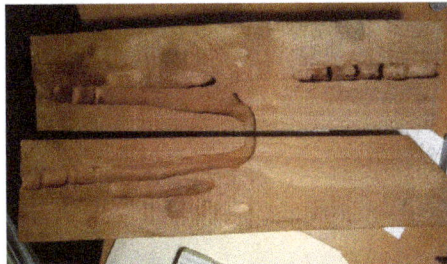

Carpenter bee nests in a cedar wood beam (sawn open)

The life cycle of a bee, be it a solitary or social species, involves the laying of an egg, the development through several moults of a legless larva, a pupation stage during which the insect undergoes complete metamorphosis, followed by the emergence of a winged adult. Most solitary bees and bumble bees in temperate climates overwinter as adults or pupae and emerge in spring when increasing numbers of flowering plants come into bloom. The males usually emerge first and search for females with which to mate. The sex of a bee is determined by whether or not the egg is fertilised; after mating, a female stores the sperm, and determines which sex is required at the time each individual egg is laid, fertilised eggs producing female offspring and unfertilised eggs, males. Tropical bees may have several generations in a year and no diapause stage.

Honeybees on brood comb with eggs and larvae in cells

The egg is generally oblong, slightly curved and tapering at one end. In the case of solitary bees, each one is laid in a cell with a supply of mixed pollen and nectar next to it. This may be rolled into a pellet or placed in a pile and is known as mass provisioning. In social species of bee there is progressive provisioning with the larva being fed regularly while it grows. The nest varies from a hole in the ground or in wood, in solitary bees, to a substantial structure with wax combs in bumblebees and honey bees.

The larvae are generally whitish grubs, roughly oval and bluntly-pointed at both ends. They have fifteen segments and spiracles in each segment for breathing. They have no legs but are able to move within the confines of the cell, helped by tubercles on their sides. They have short horns on the head, jaws for chewing their food and an appendage on either side of the mouth tipped with a bristle. There is a gland under the mouth that secretes a viscous liquid which solidifies into the silk they use to produce their cocoons. The pupa can be seen through the semi-transparent cocoon and over the course of a few days, the insect undergoes metamorphosis into the form of the adult bee. When ready to emerge, it splits its skin dorsally and climbs out of the exuviae as a winged adult and breaks out of the cell.

Flight

Honeybee in flight carrying pollen in pollen basket

In Antoine Magnan's 1934 book *Le vol des insectes*, he wrote that he and André Sainte-Laguë had applied the equations of air resistance to insects and found that their flight could not be explained by fixed-wing calculations, but that "One shouldn't be surprised that the results of the calculations don't square with reality". This has led to a common misconception that bees "violate aerodynamic theory", but in fact it merely confirms that bees do not engage in fixed-wing flight, and that their flight is explained by other mechanics, such as those used by helicopters. In 1996 it was shown that vortices created by many insects' wings helped to provide lift. High-speed cinematography and robotic mock-up of a bee wing showed that lift was generated by "the unconventional combination of short, choppy wing strokes, a rapid rotation of the wing as it flops over and reverses direction, and a very fast wing-beat frequency". Wing-beat frequency normally increases as size decreases, but as the bee's wing beat covers such a small arc, it flaps approximately 230 times per second, faster than a fruitfly (200 times per second) which is 80 times smaller.

Navigation, Communication, and Finding Food

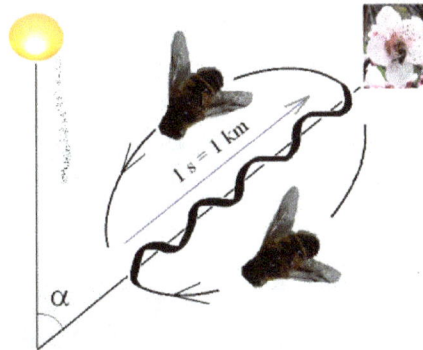

Karl von Frisch (1953) discovered that honey bee workers can navigate, indicating the range and direction to food to other workers with a waggle dance.

The ethologist Karl von Frisch studied navigation in the honey bee. He showed that honey bees communicate by the waggle dance, in which a worker indicates the location of a food source to other workers in the hive. He demonstrated that bees can recognize a desired compass direction in three different ways: by the sun, by the polarization pattern of the blue sky, and by the earth's magnetic field. He showed that the sun is the preferred or main compass; the other mechanisms are used under cloudy skies or inside a dark beehive. Bees navigate using spatial memory with a "rich, map-like organization".

Ecology

Floral Relationships

Most bees are polylectic (generalist) meaning they collect pollen from a range of flowering plants, however, some are oligoleges (specialists), in that they only gather pollen from one or a few species or genera of closely related plants. Specialist pollinators also include bee species which gather floral oils instead of pollen, and male orchid bees, which gather aromatic compounds from orchids (one of the few cases where male bees are effective pollinators). Bees are able to sense the presence of desirable flowers through ultraviolet patterning on flowers, floral odors, and even electromagnetic fields. Once landed, a bee then uses nectar quality and pollen taste to determine whether to continue visiting similar flowers.

In rare cases, a plant species may only be effectively pollinated by a single bee species, and some plants are endangered at least in part because their pollinator is also threatened. There is, however, a pronounced tendency for oligolectic bees to be associated with common, widespread plants which are visited by multiple pollinators. There are some forty oligoleges associated with the creosote bush in the arid parts of the United States southwest, for example.

As Mimics and Models

The bee-fly *Bombylius major*, a Batesian mimic of bees, taking nectar and pollinating a flower.

Bee orchid lures male bees to attempt to mate with the flower's lip, which resembles a bee perched on a pink flower.

Many bees are aposematically coloured, typically orange and black, warning of their ability to defend themselves with a powerful sting. As such they are models for Batesian mimicry by non-stinging insects such as bee-flies, robber flies and hoverflies, all of which gain a measure of protection by superficially looking and behaving like bees.

Bees are themselves Müllerian mimics of other aposematic insects with the same colour scheme, including wasps, lycid and other beetles, and many butterflies and moths (Lepidoptera) which are themselves distasteful, often through acquiring bitter and poisonous chemicals from their plant food. All the Müllerian mimics, including bees, benefit from the reduced risk of predation that results from their easily recognised warning coloration.

Bees are also mimicked by plants such as the bee orchid which imitates both the appearance and the scent of a female bee; male bees attempt to mate (pseudocopulation) with the furry lip of the flower, thus pollinating it.

As Brood Parasites

Bombus vestalis, a brood parasite of the bumblebee Bombus terrestris

Brood parasites occur in several bee families including the apid subfamily Nomadinae. Females of these bees lack pollen collecting structures (the scopa) and do not construct their own nests. They typically enter the nests of pollen collecting species, and lay their eggs in cells provisioned by the host bee. When the cuckoo bee larva hatches it consumes the host larva's pollen ball, and often the host egg also. The Arctic bee species, *Bombus hyperboreus,* in particular are an aggressive species that attack and enslave other bees of the same subgenus. However, unlike many other bee brood parasites, they have pollen baskets and often collect pollen.

In the south of Africa, hives of African honeybees (*A. mellifera scutellata*) are being destroyed by parasitic workers of the Cape honeybee, *A. m. capensis*. These lay diploid eggs ("thelytoky"), escaping normal worker policing, leading to the colony's destruction; the parasites can then move to other hives.

The cuckoo bees in the *Bombus* subgenus *Psithyrus* are closely related to, and resemble, their hosts in looks and size. This common pattern gave rise to the ecological principle "Emery's rule". Others parasitize bees in different families, like *Townsendiella,* a nomadine apid, two species of which are cleptoparasites of the dasypodaid genus *Hesperapis*, while the other species in the same genus attacks halictid bees.

Nocturnal Bees

Four bee families (Andrenidae, Colletidae, Halictidae, and Apidae) contain some species that are crepuscular. Most are tropical or subtropical, but there are some which live in arid regions at higher latitudes. These bees have greatly enlarged ocelli, which are extremely sensitive to light and dark, though incapable of forming images. Some have refracting superposition compound eyes: these combine the output of many elements of their compound eyes to provide enough light for each retinal photoreceptor. Their ability to fly by night enables them to avoid many predators, and to exploit flowers that produce nectar only or also at night.

Predators, Parasites and Pathogens

The bee-eater, Merops apiaster, specialises in feeding on bees; here a male catches a nuptial gift for his mate.

Vertebrate predators of bees include bee-eaters, shrikes and flycatchers, which make short sallies to catch insects in flight. Swifts and swallows fly almost continually, catching insects as they go. The honey buzzard attacks bees' nests and eats the larvae. The greater honeyguide interacts with humans by guiding them to the nests of wild bees. The humans break open the nests and take the honey and the bird feeds on the larvae and the wax. Among mammals, predators such as the badger dig up bumblebee nests and eat both the larvae and any stored food.

The beewolf Philanthus triangulum paralysing a bee with its sting

Specialist ambush predators of visitors to flowers include crab spiders, which wait on flowering plants for pollinating insects; predatory bugs, and praying mantises, some of which (the flower mantises of the tropics) wait motionless, aggressive mimics camouflaged as flowers. Beewolves are large wasps that habitually attack bees; the ethologist Niko Tinbergen estimated that a single colony of the beewolf *Philanthus triangulum* might kill several thousand honeybees in a day: all the prey he observed were honeybees. Other predatory insects that sometimes catch bees include robber flies and dragonflies.

Honey bees are affected by parasites including acarine and *Varroa* mites. However, some bees are believed to have a mutualistic relationship with mites.

Bees and Humans

In Mythology and Folklore

Gold plaques embossed with winged bee goddesses. Camiros, Rhodes. 7th century B.C.

Three bee maidens with the power of divination and thus speaking truth are described in Homer's *Hymn to Hermes*, and the food of the gods is "identified as honey"; the bee maidens were originally associated with Apollo, and are probably not correctly identified with the Thriae. Honey, according to a Greek myth, was discovered by a nymph called Melissa ("Bee"); and honey was offered to the Greek gods from Mycenean times. Bees were associated, too, with the Delphic oracle and the prophetess was sometimes called a bee.

The image of a community of honey bees has been used from ancient to modern times, in Aristotle and Plato; in Virgil and Seneca; in Erasmus and Shakespeare; Tolstoy, and by political and social theorists such as Bernard Mandeville and Karl Marx as a model for human society. In English folklore, bees would be told of important events in the household, in a custom known as "Telling the bees".

In Literature

Beatrix Potter's illustrated book *The Tale of Mrs Tittlemouse* (1910) features Babbity Bumble and her brood *(pictured)*.

W. B. Yeats's poem *The Lake Isle of Innisfree* (1888) contains the couplet "Nine bean rows will I have there, a hive for the honey bee, / And live alone in the bee loud glade." At the time he was living in Bedford Park in the West of London.

Beatrix Potter's illustration of Babbity Bumble in The Tale of Mrs Tittlemouse, 1910

Kit Williams' treasure hunt book *The Bee on the Comb* (1984) uses bees and beekeeping as part of its story and puzzle.

Sue Monk Kidd's *The Secret Life of Bees* (2004), and the 2009 film starring Dakota Fanning, tells the story of a girl who escapes her abusive home and finds her way to live with a family of beekeepers, the Boatwrights.

Dave Goulson's *A Sting in the Tale* (2014) describes his efforts to save bumblebees in Britain, as well as much about their biology.

The playwright Laline Paull's fantasy *The Bees* (2015) tells the tale of a hive bee named Flora 717 from hatching onwards.

The humorous 2007 animated film *Bee Movie* used Jerry Seinfeld's first script and was his first work for children; he starred as a bee named Barry B. Benson, alongside Renée Zellweger. Critics found its premise awkward and its delivery tame.

Beekeeping

A commercial beekeeper at work

Humans have kept honey bee colonies, commonly in hives, for millennia. Beekeepers collect honey, beeswax, propolis, pollen, and royal jelly from hives; bees are also kept to pollinate crops and to produce bees for sale to other beekeepers.

Depictions of humans collecting honey from wild bees date to 15,000 years ago; efforts to domesticate them are shown in Egyptian art around 4,500 years ago. Simple hives and smoke were used; jars of honey were found in the tombs of pharaohs such as Tutankhamun. From the 18th century, European understanding of the colonies and biology of bees allowed the construction of the moveable comb hive so that honey could be harvested without destroying the colony. Among Classical Era authors, beekeeping with the use of smoke is described in the *History of Animals* Book 9 (a book not written by Aristotle himself). The account mentions that bees die after stinging; that workers remove corpses from the hive, and guard it; castes including workers and non-working drones, but "kings" rather than queens; predators including toads and bee-eaters; and the waggle dance, with the "irresistible suggestion" of aroseiontai, it waggles and parakolouthousin, they watch.

Beekeeping is described in detail by Virgil in his *Eclogues*; it is also mentioned in his *Aeneid*, and in Pliny's *Natural History*.

As Commercial Pollinators

Bees play an important role in pollinating flowering plants, and are the major type of pollinator in many ecosystems that contain flowering plants. It is estimated that one third of the human food supply depends on pollination by insects, birds and bats, most of which is accomplished by bees, especially the domesticated European honey bee.

Squash bees (Apidae) are important pollinators of squashes and cucumbers.

Contract pollination has overtaken the role of honey production for beekeepers in many countries. From 1972 to 2006, feral honey bees declined dramatically in the US, and they are now almost absent. The number of colonies kept by beekeepers declined slightly, through urbanization, systematic pesticide use, tracheal and *Varroa* mites, and the closure of beekeeping businesses. In 2006 and 2007 the rate of attrition increased, and was described as colony collapse disorder. In 2010 invertebrate iridescent virus and the fungus *Nosema ceranae* were shown to be in every killed colony, and deadly in combination. Winter losses increased to about 1/3. *Varroa* mites were thought to be responsible for about half the losses.

Apart from colony collapse disorder, losses outside the US have been attributed to causes including pesticide seed dressings, such as Clothianidin, Imidacloprid and Thiamethoxam. From 2013 the European Union restricted some pesticides to stop bee populations from declining further. In 2014 the Intergovernmental Panel on Climate Change report warned that bees faced increased risk of extinction because of global warming.

However farmers have focused on alternative solutions in order to mitigate these problems. By raising native plants, they provide food for native bee pollinators like *L. vierecki* and *L. leucozonium*, leading to less reliance on honey bee populations.

Bee larvae as food in the Javanese dish botok tawon

As Food

Honey is a natural product produced by bees and stored for their own use, but its sweetness has always appealed to humans. Before domestication of bees was even attempted, humans were raiding their nests for their honey. Smoke was often used to subdue the bees and such activities are depicted in rock paintings in Spain which have been dated to 15,000 BC. Indigenous people in many countries eat insects, including consuming the larvae and pupae of bees, mostly stingless bees. They also gather "bee brood" (the larvae, pupae and surrounding cells) for consumption. In the Indonesian dish *botok tawon* from Central and East Java, bee larvae are eaten as a companion to rice, after being mixed with shredded coconut, wrapped in banana leaves, and steamed.

Honey bees are used commercially to produce honey. They also produce some substances used as dietary supplements with possible health benefits, pollen, propolis, and royal jelly, though all of these can also cause allergic reactions.

Stings

The painful stings of bees are mostly associated with the poison gland and the Dufour's gland which are abdominal exocrine glands containing various chemicals. In *Lasioglos-*

sum leucozonium, the Dufour's Gland mostly contains octadecanolide as well as some eicosanolide. There is also evidence of n-triscosane, n-heptacosane, and 22-docosanolide. However, the secretions of these glands could also be used for nest construction.

Ant

Ants are eusocial insects of the family Formicidae and, along with the related wasps and bees, belong to the order Hymenoptera. Ants evolved from wasp-like ancestors in the Cretaceous period, about 99 million years ago and diversified after the rise of flowering plants. More than 12,500 of an estimated total of 22,000 species have been classified. They are easily identified by their elbowed antennae and the distinctive node-like structure that forms their slender waists.

Ants form colonies that range in size from a few dozen predatory individuals living in small natural cavities to highly organised colonies that may occupy large territories and consist of millions of individuals. Larger colonies consist mostly of sterile, wingless females forming castes of "workers", "soldiers", or other specialised groups. Nearly all ant colonies also have some fertile males called "drones" and one or more fertile females called "queens". The colonies are described as superorganisms because the ants appear to operate as a unified entity, collectively working together to support the colony.

Ants have colonised almost every landmass on Earth. The only places lacking indigenous ants are Antarctica and a few remote or inhospitable islands. Ants thrive in most ecosystems and may form 15–25% of the terrestrial animal biomass. Their success in so many environments has been attributed to their social organisation and their ability to modify habitats, tap resources, and defend themselves. Their long co-evolution with other species has led to mimetic, commensal, parasitic, and mutualistic relationships.

Ant societies have division of labour, communication between individuals, and an ability to solve complex problems. These parallels with human societies have long been an inspiration and subject of study. Many human cultures make use of ants in cuisine, medication, and rituals. Some species are valued in their role as biological pest control agents. Their ability to exploit resources may bring ants into conflict with humans, however, as they can damage crops and invade buildings. Some species, such as the red imported fire ant (*Solenopsis invicta*), are regarded as invasive species, establishing themselves in areas where they have been introduced accidentally.

Taxonomy and Evolution

Ants fossilised in Baltic amber

The family Formicidae belongs to the order Hymenoptera, which also includes sawflies, bees, and wasps. Ants evolved from a lineage within the aculeate wasps, and a 2013 study suggests that they are a sister group of the Apoidea. In 1966, E. O. Wilson and his colleagues identified the fossil remains of an ant (*Sphecomyrma*) that lived in the Cretaceous period. The specimen, trapped in amber dating back to around 92 million years ago, has features found in some wasps, but not found in modern ants. *Sphecomyrma* possibly was a ground forager, while *Haidomyrmex* and *Haidomyrmodes*, related genera in subfamily Sphecomyrminae, are reconstructed as active arboreal predators. Older ants in the genus *Sphecomyrmodes* have been found in 99 million year-old amber from Myanmar. After the rise of flowering plants about 100 million years ago they diversified and assumed ecological dominance around 60 million years ago. Some groups, such as the Leptanillinae and Martialinae, are suggested to have diversified from early primitive ants that were likely to have been predators underneath the surface of the soil.

During the Cretaceous period, a few species of primitive ants ranged widely on the Laurasian supercontinent (the Northern Hemisphere). They were scarce in comparison to the populations of other insects, representing only about 1% of the entire insect population. Ants became dominant after adaptive radiation at the beginning of the Paleogene period. By the Oligocene and Miocene, ants had come to represent 20–40% of all insects found in major fossil deposits. Of the species that lived in the Eocene epoch, around one in 10 genera survive to the present. Genera surviving today comprise 56% of the genera in Baltic amber fossils (early Oligocene), and 92% of the genera in Dominican amber fossils (apparently early Miocene).

Termites, although sometimes called 'white ants', are not ants. They belong to the sub-order Isoptera within the order Blattodea. Termites are more closely related to cockroaches and mantids. Termites are eusocial, but differ greatly in the genetics of reproduction. The similarity of their social structure to that of ants is attributed to convergent evolution. Velvet ants look like large ants, but are wingless female wasps.

Distribution and Diversity

Region	Number of species
Neotropics	2162
Nearctic	580
Europe	180
Africa	2500
Asia	2080
Melanesia	275
Australia	985
Polynesia	42

Ants are found on all continents except Antarctica, and only a few large islands, such as Greenland, Iceland, parts of Polynesia and the Hawaiian Islands lack native ant species. Ants occupy a wide range of ecological niches and exploit many different food resources as direct or indirect herbivores, predators and scavengers. Most ant species are omnivorous generalists, but a few are specialist feeders. Their ecological dominance is demonstrated by their biomass: ants are estimated to contribute 15–20 % (on average and nearly 25% in the tropics) of terrestrial animal biomass, exceeding that of the vertebrates.

Ants range in size from 0.75 to 52 millimetres (0.030–2.0 in), the largest species being the fossil *Titanomyrma giganteum*, the queen of which was 6 centimetres (2.4 in) long with a wingspan of 15 centimetres (5.9 in). Ants vary in colour; most ants are red or black, but a few species are green and some tropical species have a metallic lustre. More than 12,000 species are currently known (with upper estimates of the potential existence of about 22,000), with the greatest diversity in the tropics. Taxonomic studies continue to resolve the classification and systematics of ants. Online databases of ant species, including AntBase and the Hymenoptera Name Server, help to keep track of the known and newly described species. The relative ease with which ants may be sampled and studied in ecosystems has made them useful as indicator species in biodiversity studies.

Morphology

Ants are distinct in their morphology from other insects in having elbowed antennae, metapleural glands, and a strong constriction of their second abdominal segment into a node-like petiole. The head, mesosoma, and metasoma are the three distinct body segments. The petiole forms a narrow waist between their mesosoma (thorax plus the first abdominal segment, which is fused to it) and gaster (abdomen less the abdominal segments in the petiole). The petiole may be formed by one or two nodes (the second alone, or the second and third abdominal segments).

Bull ant showing the powerful mandibles and the relatively large compound eyes that provide excellent vision

Like other insects, ants have an exoskeleton, an external covering that provides a protective casing around the body and a point of attachment for muscles, in contrast to the internal skeletons of humans and other vertebrates. Insects do not have lungs; oxygen and other gases, such as carbon dioxide, pass through their exoskeleton via tiny valves called spiracles. Insects also lack closed blood vessels; instead, they have a long, thin, perforated tube along the top of the body (called the "dorsal aorta") that functions like a heart, and pumps haemolymph toward the head, thus driving the circulation of the internal fluids. The nervous system consists of a ventral nerve cord that runs the length of the body, with several ganglia and branches along the way reaching into the extremities of the appendages.

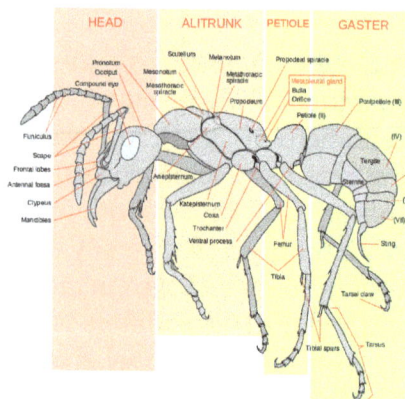

Diagram of a worker ant (Pachycondyla verenae)

Head

An ant's head contains many sensory organs. Like most insects, ants have compound eyes made from numerous tiny lenses attached together. Ant eyes are good for acute movement detection, but do not offer a high resolution image. They also have three small ocelli (simple eyes) on the top of the head that detect light levels and polarization. Compared to vertebrates, most ants have poor-to-mediocre eyesight and a few subterranean species are completely blind. However, some ants, such as Australia's bulldog ant, have excellent vision and are capable of discriminating the distance and size of objects moving nearly a metre away.

Two antennae ("feelers") are attached to the head; these organs detect chemicals, air currents, and vibrations; they also are used to transmit and receive signals through touch. The head has two strong jaws, the mandibles, used to carry food, manipulate objects, construct nests, and for defence. In some species, a small pocket (infrabuccal chamber) inside the mouth stores food, so it may be passed to other ants or their larvae.

Legs

All six legs are attached to the mesosoma ("thorax") and terminate in a hooked claw.

Wings

Only reproductive ants, queens, and males, have wings. Queens shed their wings after the nuptial flight, leaving visible stubs, a distinguishing feature of queens. In a few species, wingless queens (ergatoids) and males occur.

Metasoma

The metasoma (the "abdomen") of the ant houses important internal organs, including those of the reproductive, respiratory (tracheae), and excretory systems. Workers of many species have their egg-laying structures modified into stings that are used for subduing prey and defending their nests.

Polymorphism

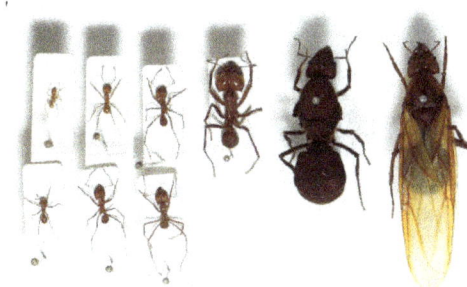

Seven Leafcutter ant workers of various castes (left) and two Queens (right)

In the colonies of a few ant species, there are physical castes—workers in distinct size-classes, called minor, median, and major workers. Often, the larger ants have disproportionately larger heads, and correspondingly stronger mandibles. Such individuals are sometimes called "soldier" ants because their stronger mandibles make them more effective in fighting, although they still are workers and their "duties" typically do not vary greatly from the minor or median workers. In a few species, the median workers are absent, creating a sharp divide between the minors and majors. Weaver ants, for example, have a distinct bimodal size distribution. Some other species show continuous variation in the size of workers. The smallest and largest workers in *Pheidologeton diversus* show nearly a 500-fold difference in their dry-weights. Workers cannot mate; however, because of the haplodiploid sex-determination system in ants, workers of a number of species can lay unfertilised eggs that become fully fertile, haploid males. The role of workers may change with their age and in some species, such as honeypot ants, young workers are fed until their gasters are distended, and act as living food storage vessels. These food storage workers are called *repletes*. For instance, these replete workers develop in the North American honeypot ant *Myrmecocystus mexicanus*. Usually the largest workers in the colony develop into repletes; and, if repletes are removed from the colony, other workers become repletes, demonstrating the flexibility of this particular polymorphism. This polymorphism in morphology and behaviour of workers initially was thought to be determined by environmental factors such as nutrition and hormones that led to different developmental paths; however, genetic differences between worker castes have been noted in *Acromyrmex* sp. These polymorphisms are caused by relatively small genetic changes; differences in a single gene of *Solenopsis invicta* can decide whether the colony will have single or multiple queens. The Australian jack jumper ant (*Myrmecia pilosula*) has only a single pair of chromosomes (with the males having just one chromosome as they are haploid), the lowest number known for any animal, making it an interesting subject for studies in the genetics and developmental biology of social insects.

Life Cycle

Meat eater ant nest during swarming

The life of an ant starts from an egg. If the egg is fertilised, the progeny will be female diploid; if not, it will be male haploid. Ants develop by complete metamorphosis with the larva stages passing through a pupal stage before emerging as an adult. The larva is largely immobile and is fed and cared for by workers. Food is given to the larvae by trophallaxis, a process in which an ant regurgitates liquid food held in its crop. This is also how adults share food, stored in the "social stomach". Larvae, especially in the later stages, may also be provided solid food, such as trophic eggs, pieces of prey, and seeds brought by workers.

The larvae grow through a series of four or five moults and enter the pupal stage. The pupa has the appendages free and not fused to the body as in a butterfly pupa. The differentiation into queens and workers (which are both female), and different castes of workers, is influenced in some species by the nutrition the larvae obtain. Genetic influences and the control of gene expression by the developmental environment are complex and the determination of caste continues to be a subject of research. Winged male ants, called drones, emerge from pupae along with the usually winged breeding females. Some species, such as army ants, have wingless queens. Larvae and pupae need to be kept at fairly constant temperatures to ensure proper development, and so often, are moved around among the various brood chambers within the colony.

A new worker spends the first few days of its adult life caring for the queen and young. She then graduates to digging and other nest work, and later to defending the nest and foraging. These changes are sometimes fairly sudden, and define what are called temporal castes. An explanation for the sequence is suggested by the high casualties involved in foraging, making it an acceptable risk only for ants who are older and are likely to die soon of natural causes.

Ant colonies can be long-lived. The queens can live for up to 30 years, and workers live from 1 to 3 years. Males, however, are more transitory, being quite short-lived and surviving for only a few weeks. Ant queens are estimated to live 100 times as long as solitary insects of a similar size.

Ants are active all year long in the tropics, but, in cooler regions, they survive the winter in a state of dormancy or inactivity. The forms of inactivity are varied and some temperate species have larvae going into the inactive state, (diapause), while in others, the adults alone pass the winter in a state of reduced activity.

Reproduction

Ants Mating

A wide range of reproductive strategies have been noted in ant species. Females of many species are known to be capable of reproducing asexually through thelytokous parthenogenesis. Secretions from the male accessory glands in some species can plug the female genital opening and prevent females from re-mating. Most ant species have a system in which only the queen and breeding females have the ability to mate. Contrary to popular belief, some ant nests have multiple queens, while others may exist without queens. Workers with the ability to reproduce are called "gamergates" and colonies that lack queens are then called gamergate colonies; colonies with queens are said to be queen-right.

Drones can also mate with existing queens by entering a foreign colony. When the drone is initially attacked by the workers, it releases a mating pheromone. If recognized as a mate, it will be carried to the queen to mate. Males may also patrol the nest and fight others by grabbing them with their mandibles, piercing their exoskeleton and then marking them with a pheromone. The marked male is interpreted as an invader by worker ants and is killed.

Most ants are univoltine, producing a new generation each year. During the species-specific breeding period, new reproductives, females, and winged males leave the colony in what is called a nuptial flight. The nuptial flight usually takes place in the late spring or early summer when the weather is hot and humid. Heat makes flying easier and freshly fallen rain makes the ground softer for mated queens to dig nests. Males typically take flight before the females. Males then use visual cues to find a common mating ground, for example, a landmark such as a pine tree to which other males in the area converge. Males secrete a mating pheromone that females follow. Males will mount females in the air, but the actual mating process usually takes place on the ground. Females of some species mate with just one male but in others they may mate with as many as ten or more different males, storing the sperm in their spermathecae.

Fertilised meat-eater ant queen beginning to dig a new colony

Mated females then seek a suitable place to begin a colony. There, they break off their wings and begin to lay and care for eggs. The females can selectively fertilise future eggs with the sperm stored or lay unfertilized haploid eggs to produce workers. The first workers to hatch are weak and smaller than later workers, but they begin to serve the colony immediately. They enlarge the nest, forage for food, and care for the other eggs. Species that have multiple queens may have a queen leaving the nest along with some workers to found a colony at a new site, a process akin to swarming in honeybees.

Behaviour and Ecology

Communication

Two Camponotus sericeus workers communicating through touch and pheromones

Ants communicate with each other using pheromones, sounds, and touch. The use of pheromones as chemical signals is more developed in ants, such as the red harvester ant, than in other hymenopteran groups. Like other insects, ants perceive smells with their long, thin, and mobile antennae. The paired antennae provide information about the direction and intensity of scents. Since most ants live on the ground, they use the soil surface to leave pheromone trails that may be followed by other ants. In species that forage in groups, a forager that finds food marks a trail on the way back to the colony; this trail is followed by other ants, these ants then reinforce the trail when they head back with food to the colony. When the food source is exhausted, no new trails are marked by returning ants and the scent slowly dissipates. This behaviour helps ants deal with changes in their environment. For instance, when an established path to a food source is blocked by an obstacle, the foragers leave the path to explore new routes. If an ant is successful, it leaves a new trail marking the shortest route on its return. Successful trails are followed by more ants, reinforcing better routes and gradually identifying the best path.

Ants use pheromones for more than just making trails. A crushed ant emits an alarm pheromone that sends nearby ants into an attack frenzy and attracts more ants from farther away. Several ant species even use "propaganda pheromones" to confuse enemy ants and make them fight among themselves. Pheromones are produced by a wide range of structures including Dufour's glands, poison glands and glands on the hindgut, pygidium, rectum, sternum, and hind tibia. Pheromones also are exchanged, mixed with food, and passed by trophallaxis, transferring information within the colony. This allows other ants to detect what task group (*e.g.*, foraging or nest maintenance) other colony members belong to. In ant species with queen castes, when the dominant queen stops producing a specific pheromone, workers begin to raise new queens in the colony.

Some ants produce sounds by stridulation, using the gaster segments and their mandibles. Sounds may be used to communicate with colony members or with other species.

Defence

A *Plectroctena* sp. attacks another of its kind to protect its territory

Ants attack and defend themselves by biting and, in many species, by stinging, often injecting or spraying chemicals, such as formic acid in the case of formicine ants, alkaloids and piperidines in fire ants, and a variety of protein components in other ants. Bullet ants (*Paraponera*), located in Central and South America, are considered to have the most painful sting of any insect, although it is usually not fatal to humans. This sting is given the highest rating on the Schmidt Sting Pain Index.

The sting of jack jumper ants can be fatal, and an antivenom has been developed for it.

Fire ants, *Solenopsis* spp., are unique in having a venom sac containing piperidine alkaloids. Their stings are painful and can be dangerous to hypersensitive people.

A weaver ant in fighting position, mandibles wide open

Trap-jaw ants of the genus *Odontomachus* are equipped with mandibles called trap-jaws, which snap shut faster than any other predatory appendages within the animal kingdom. One study of *Odontomachus bauri* recorded peak speeds of between 126 and 230 km/h (78 and 143 mph), with the jaws closing within 130 microseconds on average. The ants were also observed to use their jaws as a catapult to eject intruders or fling themselves backward to escape a threat. Before striking, the ant opens its mandibles extremely widely and locks them in this position by an internal mechanism. Energy is stored in a thick band of muscle and explosively released when triggered by the stimulation of sensory organs resembling hairs on the inside of the mandibles. The mandibles also permit slow and fine movements for other tasks. Trap-jaws also are seen in the following genera: *Anochetus*, *Orectognathus*, and *Strumigenys*, plus some members of the Dacetini tribe, which are viewed as examples of convergent evolution.

A Malaysian species of ant in the *Camponotus cylindricus* group has enlarged mandibular glands that extend into their gaster. When disturbed, workers rupture the membrane of the gaster, causing a burst of secretions containing acetophenones and other chemicals that immobilise small insect attackers. The worker subsequently dies.

Suicidal defences by workers are also noted in a Brazilian ant, *Forelius pusillus*, where a small group of ants leaves the security of the nest after sealing the entrance from the outside each evening.

Ant mound holes prevent water from entering the nest during rain.

In addition to defence against predators, ants need to protect their colonies from pathogens. Some worker ants maintain the hygiene of the colony and their activities include undertaking or *necrophory*, the disposal of dead nest-mates. Oleic acid has been identified as the compound released from dead ants that triggers necrophoric behaviour in *Atta mexicana* while workers of *Linepithema humile* react to the absence of characteristic chemicals (dolichodial and iridomyrmecin) present on the cuticle of their living nestmates to trigger similar behaviour.

Nests may be protected from physical threats such as flooding and overheating by elaborate nest architecture. Workers of *Cataulacus muticus*, an arboreal species that lives in plant hollows, respond to flooding by drinking water inside the nest, and excreting it outside. *Camponotus anderseni*, which nests in the cavities of wood in mangrove habitats, deals with submergence under water by switching to anaerobic respiration.

Learning

Many animals can learn behaviours by imitation, but ants may be the only group apart from mammals where interactive teaching has been observed. A knowledgeable forager of *Temnothorax albipennis* will lead a naive nest-mate to newly discovered food by the process of tandem running. The follower obtains knowledge through its leading tutor. The leader is acutely sensitive to the progress of the follower and slows down when the follower lags and speeds up when the follower gets too close.

Controlled experiments with colonies of *Cerapachys biroi* suggest that an individual may choose nest roles based on her previous experience. An entire generation of identical workers was divided into two groups whose outcome in food foraging was controlled. One group was continually rewarded with prey, while it was made certain that the other failed. As a result, members of the successful group intensified their foraging attempts while the unsuccessful group ventured out fewer and fewer times. A month later, the successful foragers continued in their role while the others had moved to specialise in brood care.

Nest Construction

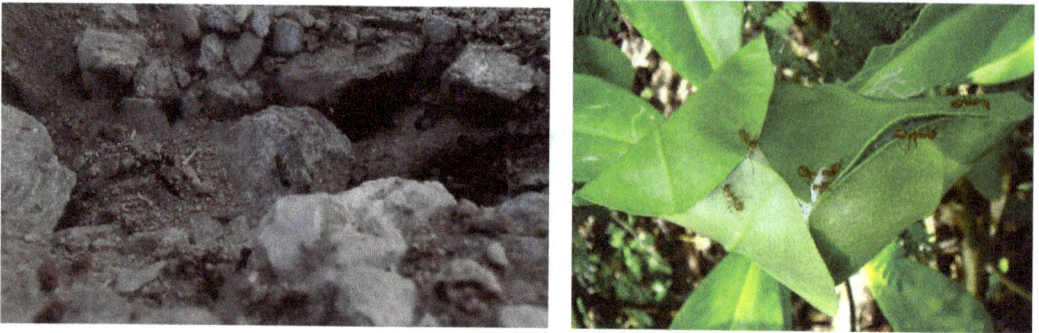

Leaf nest of weaver ants, Pamalican, Philippines

Complex nests are built by many ant species, but other species are nomadic and do not build permanent structures. Ants may form subterranean nests or build them on trees. These nests may be found in the ground, under stones or logs, inside logs, hollow stems, or even acorns. The materials used for construction include soil and plant matter, and ants carefully select their nest sites; *Temnothorax albipennis* will avoid sites with dead ants, as these may indicate the presence of pests or disease. They are quick to abandon established nests at the first sign of threats.

The army ants of South America, such as the *Eciton burchellii* species, and the driver ants of Africa do not build permanent nests, but instead, alternate between nomadism and stages where the workers form a temporary nest (bivouac) from their own bodies, by holding each other together.

Weaver ant (*Oecophylla* spp.) workers build nests in trees by attaching leaves together, first pulling them together with bridges of workers and then inducing their larvae to produce silk as they are moved along the leaf edges. Similar forms of nest construction are seen in some species of *Polyrhachis*.

Formica polyctena, among other ant species, constructs nests that maintain a relatively constant interior temperature that aids in the development of larvae. The ants maintain the nest temperature by choosing the location, nest materials, controlling ventilation and maintaining the heat from solar radiation, worker activity and metabolism, and in some moist nests, microbial activity in the nest materials.

Some ant species, such as those that use natural cavities, can be opportunistic and make use of the controlled micro-climate provided inside human dwellings and other artificial structures to house their colonies and nest structures.

Cultivation of Food

Myrmecocystus, honeypot ants, store food to prevent colony famine

Most ants are generalist predators, scavengers, and indirect herbivores, but a few have evolved specialised ways of obtaining nutrition. It is believed that many ant species that engage in indirect herbivory rely on specialized symbiosis with their gut microbes to upgrade the nutritional value of the food they collect and allow them to survive in nitrogen poor regions, such as rainforrest canopies. Leafcutter ants (*Atta* and *Acromyrmex*) feed exclusively on a fungus that grows only within their colonies. They continually collect leaves which are taken to the colony, cut into tiny pieces and placed in fungal gardens. Workers specialise in related tasks according to their sizes. The largest ants cut stalks, smaller workers chew the leaves and the smallest tend the fungus. Leafcutter ants are sensitive enough to recognise the reaction of the fungus to different plant material, apparently detecting chemical signals from the fungus. If a particular type of leaf is found to be toxic to the fungus, the colony will no longer collect it. The ants feed on structures produced by the fungi called *gongylidia*. Symbiotic bacteria on the exterior surface of the ants produce antibiotics that kill bacteria introduced into the nest that may harm the fungi.

Navigation

An ant trail

Foraging ants travel distances of up to 200 metres (700 ft) from their nest and scent trails allow them to find their way back even in the dark. In hot and arid regions, day-foraging ants face death by desiccation, so the ability to find the shortest route back to the nest reduces that risk. Diurnal desert ants of the genus *Cataglyphis* such as the Sahara desert ant navigate by keeping track of direction as well as distance travelled. Distances travelled are measured using an internal pedometer that keeps count of the steps taken and also by evaluating the movement of objects in their visual field (optical flow). Directions are measured using the position of the sun. They integrate this information to find the shortest route back to their nest. Like all ants, they can also make use of visual landmarks when available as well as olfactory and tactile cues to navigate. Some species of ant are able to use the Earth's magnetic field for navigation. The compound eyes of ants have specialised cells that detect polarised light from the Sun, which is used to determine direction. These polarization detectors are sensitive in the ultraviolet region of the light spectrum. In some army ant species, a group of foragers who become separated from the main column may sometimes turn back on themselves and form a circular ant mill. The workers may then run around continuously until they die of exhaustion.

Locomotion

The female worker ants do not have wings and reproductive females lose their wings after their mating flights in order to begin their colonies. Therefore, unlike their wasp ancestors, most ants travel by walking. Some species are capable of leaping. For example, Jerdon's jumping ant (*Harpegnathos saltator*) is able to jump by synchronising the action of its mid and hind pairs of legs. There are several species of gliding ant including *Cephalotes atratus*; this may be a common trait among most arboreal ants. Ants with this ability are able to control the direction of their descent while falling.

Other species of ants can form chains to bridge gaps over water, underground, or through spaces in vegetation. Some species also form floating rafts that help them survive floods. These rafts may also have a role in allowing ants to colonise islands. *Polyrhachis sokolova*, a species of ant found in Australian mangrove swamps, can swim and live in underwater nests. Since they lack gills, they go to trapped pockets of air in the submerged nests to breathe.

Cooperation and Competition

Meat-eater ants feeding on a cicada, social ants cooperate and collectively gather food

Not all ants have the same kind of societies. The Australian bulldog ants are among the biggest and most basal of ants. Like virtually all ants, they are eusocial, but their social behaviour is poorly developed compared to other species. Each individual hunts alone, using her large eyes instead of chemical senses to find prey.

Some species (such as *Tetramorium caespitum*) attack and take over neighbouring ant colonies. Others are less expansionist, but just as aggressive; they invade colonies to steal eggs or larvae, which they either eat or raise as workers or slaves. Extreme specialists among these slave-raiding ants, such as the Amazon ants, are incapable of feeding themselves and need captured workers to survive. Captured workers of the enslaved species *Temnothorax* have evolved a counter strategy, destroying just the female pupae of the slave-making *Protomognathus americanus*, but sparing the males (who don't take part in slave-raiding as adults).

A worker Harpegnathos saltator (a jumping ant) engaged in battle with a rival colony's queen

Ants identify kin and nestmates through their scent, which comes from hydrocarbon-laced secretions that coat their exoskeletons. If an ant is separated from its original colony, it will eventually lose the colony scent. Any ant that enters a colony without a matching scent will be attacked. Also, the reason why two separate colonies of ants will attack each other even if they are of the same species is because the genes responsible for pheromone production are different between them. The Argentine ant, however, does not have this characteristic, due to lack of genetic diversity, and has become a global pest because of it.

Parasitic ant species enter the colonies of host ants and establish themselves as social parasites; species such as *Strumigenys xenos* are entirely parasitic and do not have workers, but instead, rely on the food gathered by their *Strumigenys perplexa* hosts. This form of parasitism is seen across many ant genera, but the parasitic ant is usually a species that is closely related to its host. A variety of methods are employed to enter the nest of the host ant. A parasitic queen may enter the host nest before the first brood has hatched, establishing herself prior to development of a colony scent. Other species use pheromones to confuse the host ants or to trick them into carrying the parasitic queen into the nest. Some simply fight their way into the nest.

A conflict between the sexes of a species is seen in some species of ants with these reproducers apparently competing to produce offspring that are as closely related to them as possible. The most extreme form involves the production of clonal offspring. An extreme of sexual conflict is seen in *Wasmannia auropunctata*, where the queens produce diploid daughters by thelytokous parthenogenesis and males produce clones by a process whereby a diploid egg loses its maternal contribution to produce haploid males who are clones of the father.

Relationships with Other Organisms

The spider Myrmarachne plataleoides (female shown) mimics weaver ants to avoid predators.

Ants form symbiotic associations with a range of species, including other ant species, other insects, plants, and fungi. They also are preyed on by many animals and even certain fungi. Some arthropod species spend part of their lives within ant nests, either preying on ants, their larvae, and eggs, consuming the food stores of the ants, or avoiding predators. These inquilines may bear a close resemblance to ants. The nature of this ant mimicry (myrmecomorphy) varies, with some cases involving Batesian mimicry, where the mimic reduces the risk of predation. Others show Wasmannian mimicry, a form of mimicry seen only in inquilines.

An ant collects honeydew from an aphid

Aphids and other hemipteran insects secrete a sweet liquid called honeydew, when they feed on plant sap. The sugars in honeydew are a high-energy food source, which many ant species collect. In some cases, the aphids secrete the honeydew in response to ants tapping them with their antennae. The ants in turn keep predators away from the aphids and will move them from one feeding location to another. When migrating to a new area, many colonies will take the aphids with them, to ensure a continued supply of honeydew. Ants also tend mealybugs to harvest their honeydew. Mealybugs may become a serious pest of pineapples if ants are present to protect mealybugs from their natural enemies.

Myrmecophilous (ant-loving) caterpillars of the butterfly family Lycaenidae (e.g., blues, coppers, or hairstreaks) are herded by the ants, led to feeding areas in the daytime, and brought inside the ants' nest at night. The caterpillars have a gland which secretes honeydew when the ants massage them. Some caterpillars produce vibrations and sounds that are perceived by the ants. Other caterpillars have evolved from ant-loving to ant-eating: these myrmecophagous caterpillars secrete a pheromone that makes the ants act as if the caterpillar is one of their own larvae. The caterpillar is then taken into the ant nest where it feeds on the ant larvae. Fungus-growing ants that make up the tribe Attini, including leafcutter ants, cultivate certain species of fungus in the *Leucoagaricus* or *Leucocoprinus* genera of the Agaricaceae family. In this ant-fungus mutualism, both species depend on each other for survival. The ant *Allomerus decemarticulatus* has evolved a three-way association with the host plant, *Hirtella physophora* (Chrysobalanaceae), and a sticky fungus which is used to trap their insect prey.

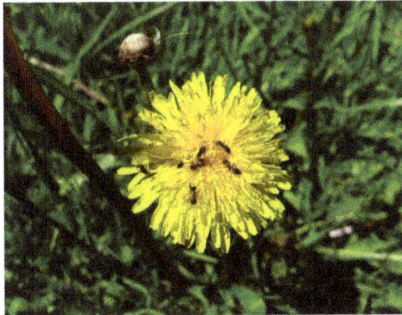

Ants may obtain nectar from flowers such as the dandelion but are only rarely known to pollinate flowers.

Lemon ants make devil's gardens by killing surrounding plants with their stings and leaving a pure patch of lemon ant trees, (*Duroia hirsuta*). This modification of the forest provides the ants with more nesting sites inside the stems of the *Duroia* trees. Although some ants obtain nectar from flowers, pollination by ants is somewhat rare. Some plants have special nectar exuding structures, extrafloral nectaries, that provide food for ants, which in turn protect the plant from more damaging herbivorous insects. Species such as the bullhorn acacia (*Acacia cornigera*) in Central America have hollow thorns that house colonies of stinging ants (*Pseudomyrmex ferruginea*) who defend the tree against insects, browsing mammals, and epiphytic vines. Isotopic labelling studies suggest that plants also obtain nitrogen from the ants. In return, the ants obtain food from protein- and lipid-rich Beltian bodies. Another example of this type of ectosymbiosis comes from the *Macaranga* tree, which has stems adapted to house colonies of *Crematogaster* ants.

Many tropical tree species have seeds that are dispersed by ants. Seed dispersal by ants or myrmecochory is widespread and new estimates suggest that nearly 9% of all plant species may have such ant associations. Some plants in fire-prone grassland systems are particularly dependent on ants for their survival and dispersal as the seeds are transported to safety below the ground. Many ant-dispersed seeds have special external structures, elaiosomes, that are sought after by ants as food.

A convergence, possibly a form of mimicry, is seen in the eggs of stick insects. They have an edible elaiosome-like structure and are taken into the ant nest where the young hatch.

A meat ant tending a common leafhopper nymph

Most ants are predatory and some prey on and obtain food from other social insects including other ants. Some species specialise in preying on termites (*Megaponera* and *Termitopone*) while a few Cerapachyinae prey on other ants. Some termites, including *Nasutitermes corniger*, form associations with certain ant species to keep away predatory ant species. The tropical wasp *Mischocyttarus drewseni* coats the pedicel of its nest with an ant-repellent chemical. It is suggested that many tropical wasps may build their nests in trees and cover them to protect themselves from ants. Other wasps such as *A. multipicta* defend against ants by blasting them off the nest with bursts of wing buzzing. Stingless bees (*Trigona* and *Melipona*) use chemical defences against ants. Certain species of ants have the power to drive certain wasps, such as *Polybia occidentalis* to extinction if they attack more than once and the wasps cannot keep up with rebuilding their nest.

Flies in the Old World genus *Bengalia* (Calliphoridae) prey on ants and are kleptoparasites, snatching prey or brood from the mandibles of adult ants. Wingless and legless females of the Malaysian phorid fly (*Vestigipoda myrmolarvoidea*) live in the nests of ants of the genus *Aenictus* and are cared for by the ants.

Fungi in the genera *Cordyceps* and *Ophiocordyceps* infect ants. Ants react to their infection by climbing up plants and sinking their mandibles into plant tissue. The fungus kills the ants, grows on their remains, and produces a fruiting body. It appears that the fungus alters the behaviour of the ant to help disperse its spores in a microhabitat that best suits the fungus. Strepsipteran parasites also manipulate their ant host to climb grass stems, to help the parasite find mates.

A nematode (*Myrmeconema neotropicum*) that infects canopy ants (*Cephalotes atratus*) causes the black-coloured gasters of workers to turn red. The parasite also alters the behaviour of the ant, causing them to carry their gasters high. The conspicuous red gasters are mistaken by birds for ripe fruits, such as *Hyeronima alchorneoides*, and eaten. The droppings of the bird are collected by other ants and fed to their young, leading to further spread of the nematode.

Spiders sometimes feed on ants

South American poison dart frogs in the genus *Dendrobates* feed mainly on ants, and the toxins in their skin may come from the ants.

Army ants forage in a wide roving column, attacking any animals in that path that are unable to escape. In Central and South America, *Eciton burchellii* is the swarming ant most commonly attended by "ant-following" birds such as antbirds and woodcreepers. This behaviour was once considered mutualistic, but later studies found the birds to be parasitic. Direct kleptoparasitism (birds stealing food from the ants' grasp) is rare and has been noted in Inca doves which pick seeds at nest entrances as they are being transported by species of *Pogonomyrmex*. Birds that follow ants eat many prey insects and thus decrease the foraging success of ants. Birds indulge in a peculiar behaviour called anting that, as yet, is not fully understood. Here birds rest on ant nests, or pick and drop ants onto their wings and feathers; this may be a means to remove ectoparasites from the birds.

Anteaters, aardvarks, pangolins, echidnas and numbats have special adaptations for living on a diet of ants. These adaptations include long, sticky tongues to capture ants and strong claws to break into ant nests. Brown bears (*Ursus arctos*) have been found to feed on ants. About 12%, 16%, and 4% of their faecal volume in spring, summer, and autumn, respectively, is composed of ants.

Relationship with Humans

Ants perform many ecological roles that are beneficial to humans, including the suppression of pest populations and aeration of the soil. The use of weaver ants in citrus cultivation in southern China is considered one of the oldest known applications of biological control. On the other hand, ants may become nuisances when they invade buildings, or cause economic losses.

Weaver ants are used as a biological control for citrus cultivation in southern China

In some parts of the world (mainly Africa and South America), large ants, especially army ants, are used as surgical sutures. The wound is pressed together and ants are applied along it. The ant seizes the edges of the wound in its mandibles and locks in place. The body is then cut off and the head and mandibles remain in place to close the wound. The large heads of the soldiers of the leafcutting ant *Atta cephalotes* are also used by native surgeons in closing wounds.

Some ants have toxic venom and are of medical importance. The species include *Paraponera clavata* (tocandira) and *Dinoponera* spp. (false tocandiras) of South America and the *Myrmecia* ants of Australia.

In South Africa, ants are used to help harvest rooibos (*Aspalathus linearis*), which are small seeds used to make a herbal tea. The plant disperses its seeds widely, making manual collection difficult. Black ants collect and store these and other seeds in their nest, where humans can gather them *en masse*. Up to half a pound (200 g) of seeds may be collected from one ant-heap.

Although most ants survive attempts by humans to eradicate them, a few are highly endangered. These tend to be island species that have evolved specialized traits and risk being displaced by introduced ant species. Examples include the critically endangered Sri Lankan relict ant (*Aneuretus simoni*) and *Adetomyrma venatrix* of Madagascar.

It has been estimated by E.O. Wilson that the total number of individual ants alive in the world at any one time is between one and ten quadrillion (short scale) (i.e. between 10^{15} and 10^{16}). According to this estimate, the total biomass of all the ants in the world is approximately equal to the total biomass of the entire human race. Also, according to this estimate, there are approximately 1 million ants for every human on Earth.

As Food

Ants and their larvae are eaten in different parts of the world. The eggs of two species of ants are used in Mexican *escamoles*. They are considered a form of insect caviar and can sell for as much as US$40 per pound ($90/kg) because they are seasonal and hard to find. In the Colombian department of Santander, *hormigas culonas* (roughly interpreted as "large-bottomed ants") *Atta laevigata* are toasted alive and eaten.

Ants and their larvae are eaten in different parts of the world. The eggs of two species of ants are used in Mexican *escamoles*. They are considered a form of insect caviar and can sell for as much as US$40 per pound ($90/kg) because they are seasonal and hard to find. In the Colombian department of Santander, *hormigas culonas* (roughly interpreted as "large-bottomed ants") *Atta laevigata* are toasted alive and eaten.

Ants and their larvae are eaten in different parts of the world. The eggs of two species of ants are used in Mexican *escamoles*. They are considered a form of insect caviar and can sell for as much as US$40 per pound ($90/kg) because they are seasonal and hard to find. In the Colombian department of Santander, *hormigas culonas* (roughly interpreted as "large-bottomed ants") *Atta laevigata* are toasted alive and eaten.

Roasted ants in Colombia

In areas of India, and throughout Burma and Thailand, a paste of the green weaver ant (*Oecophylla smaragdina*) is served as a condiment with curry. Weaver ant eggs and larvae, as well as the ants, may be used in a Thai salad, *yam* in a dish called *yam khai mot daeng* or red ant egg salad, a dish that comes from the Issan or north-eastern region of Thailand. Saville-Kent, in the *Naturalist in Australia* wrote "Beauty, in the case of the green ant, is more than skin-deep. Their attractive, almost sweetmeat-like translucency possibly invited the first essays at their consumption by the human species". Mashed up in water, after the manner of lemon squash, "these ants form a pleasant acid drink which is held in high favor by the natives of North Queensland, and is even appreciated by many European palates".

Ant larvae for sale in Isaan, Thailand

In his *First Summer in the Sierra*, John Muir notes that the Digger Indians of California ate the tickling, acid gasters of the large jet-black carpenter ants. The Mexican Indians eat the replete workers, or living honey-pots, of the honey ant (*Myrmecocystus*).

As Pests

The tiny pharaoh ant is a major pest in hospitals and office blocks; it can make nests between sheets of paper

Some ant species are considered as pests, primarily those that occur in human habitations, where their presence is often problematic. For example, the presence of ants would be undesirable in sterile places such as hospitals or kitchens. Some species or genera commonly categorized as pests include the Argentine ant, pavement ant, yellow crazy ant, banded sugar ant, Pharaoh ant, carpenter ants, odorous house ant, red imported fire ant, and European fire ant. Some ants will raid stored food, others may damage indoor structures, some can damage agricultural crops directly (or by aiding sucking pests), and some will sting or bite. The adaptive nature of ant colonies make it nearly impossible to eliminate entire colonies and most pest management practices aim to control local populations and tend to be temporary solutions. Ant populations are managed by a combination of approaches that make use of chemical, biological and physical methods. Chemical methods include the use of insecticidal bait which is gathered by ants as food and brought back to the nest where the poison is inadvertently spread to other colony members through trophallaxis. Management is based on the species and techniques can vary according to the location and circumstance.

In Science and Technology

Camponotus nearcticus workers travelling between two formicaria through connector tubing

Observed by humans since the dawn of history, the behaviour of ants has been documented and the subject of early writings and fables passed from one century to another. Those using scientific methods, myrmecologists, study ants in the laboratory and in their natural conditions. Their complex and variable social structures have made ants ideal model organisms. Ultraviolet vision was first discovered in ants by Sir John Lubbock in 1881. Studies on ants have tested hypotheses in ecology and sociobiology, and have been particularly important in examining the predictions of theories of kin selection and evolutionarily stable strategies. Ant colonies may be studied by rearing or temporarily maintaining them in *formicaria*, specially constructed glass framed enclosures. Individuals may be tracked for study by marking them with dots of colours.

The successful techniques used by ant colonies have been studied in computer science and robotics to produce distributed and fault-tolerant systems for solving problems, for example Ant colony optimization and Ant robotics. This area of biomimetics has led to studies of ant locomotion, search engines that make use of "foraging trails", fault-tolerant storage, and networking algorithms.

In Culture

Aesop's ants: picture by Milo Winter, 1888–1956

Anthropomorphised ants have often been used in fables and children's stories to represent industriousness and cooperative effort. They also are mentioned in religious texts. In the Book of Proverbs in the Bible, ants are held up as a good example for humans for their hard work and cooperation. Aesop did the same in his fable The Ant and the Grasshopper. In the Quran, Sulayman is said to have heard and understood an ant warning other ants to return home to avoid being accidentally crushed by Sulayman and his marching army. In parts of Africa, ants are considered to be the messengers of the deities. Some Native American mythology, such as the Hopi mythology, considers ants as the very first animals. Ant bites are often said to have curative properties. The sting of some species of *Pseudomyrmex* is claimed to give fever relief. Ant bites are used in the initiation ceremonies of some Amazon Indian cultures as a test of endurance.

Ant society has always fascinated humans and has been written about both humorously and seriously. Mark Twain wrote about ants in his 1880 book *A Tramp Abroad*. Some modern authors have used the example of the ants to comment on the relationship between society and the individual. Examples are Robert Frost in his poem "Departmental" and T. H. White in his fantasy novel *The Once and Future King*. The plot in French entomologist and writer Bernard Werber's *Les Fourmis* science-fiction trilogy is divided between the worlds of ants and humans; ants and their behaviour is described using contemporary scientific knowledge. H.G. Wells wrote about intelligent ants destroying human settlements in Brazil and threatening human civilization in his 1905 science-fiction short story, *The Empire of the Ants*. In more recent times, animated cartoons and 3-D animated films featuring ants have been produced including *Antz, A Bug's Life, The Ant Bully, The Ant and the Aardvark, Ferdy the Ant* and *Atom Ant*. Renowned myrmecologist E. O. Wilson wrote a short story, "Trailhead" in 2010 for *The New Yorker* magazine, which describes the life and death of an ant-queen and the rise and fall of her colony, from an ants' point of view.

From the late 1950s through the late 1970s, ant farms were popular educational children's toys in the United States. Later versions use transparent gel instead of soil, allowing greater visibility. In the early 1990s, the video game SimAnt, which simulated an ant colony, won the 1992 Codie award for "Best Simulation Program".

Ants also are quite popular inspiration for many science-fiction insectoids, such as the Formics of *Ender's Game*, the Bugs of *Starship Troopers*, the giant ants in the films *Them!* and *Empire of the Ants*, Marvel Comics' super hero Ant-Man, and ants mutated into super-intelligence in *Phase IV*. In computer strategy games, ant-based species often benefit from increased production rates due to their single-minded focus, such as the Klackons in the *Master of Orion* series of games or the ChCht in *Deadlock II*. These characters are often credited with a hive mind, a common misconception about ant colonies.

Grasshopper

Grasshoppers are insects of the order Orthoptera, suborder Caelifera. They are sometimes referred to as short-horned grasshoppers to distinguish them from the katydids (bush crickets) which have much longer antennae. They are typically ground-dwelling insects with powerful hind legs which enable them to escape from threats by leaping vigorously. They are hemimetabolous insects (do not undergo complete metamorphosis) which hatch from an egg into a nymph or "hopper" which undergoes five moults, becoming more similar to the adult insect at each developmental stage. At high population densities and under certain environmental conditions, some grasshopper species can change colour and behaviour and form swarms. Under these circumstances they are known as locusts.

Grasshoppers are plant-eaters, sometimes becoming serious pests of cereals, vegetables and pasture, especially when they swarm in their millions as locusts and destroy crops over wide areas. They protect themselves from predators by camouflage; when detected, many species attempt to startle the predator with a brilliantly-coloured wing-flash while jumping and (if adult) launching themselves into the air, usually flying for only a short distance. Other species such as the rainbow grasshopper have warning coloration which deters predators. Grasshoppers are affected by parasites and various diseases, and many predatory creatures feed on both nymphs and adults. The eggs are the subject of attack by parasitoids and predators.

Grasshoppers have had a long relationship with humans. Swarms of locusts have had dramatic effects that have changed the course of history, and even in smaller numbers grasshoppers can be serious pests. They are eaten as food and also feature in art, symbolism and literature.

Characteristics

Grasshoppers have the typical insect body plan of head, thorax and abdomen. The head is held vertically, at an angle to the body with the mouth at the bottom. It bears a large pair of compound eyes which give all-round vision, three simple eyes which can detect light and dark and a pair of thread-like antennae which are sensitive to touch and smell. The downward-directed mouthparts are modified for chewing and there are two sensory palps in front of the jaws.

The thorax and abdomen are segmented and have a rigid cuticle made up of overlapping plates composed of chitin. The three fused thoracic segments bear three pairs of legs and two pairs of wings. The forewings, known as tegmina, are narrow and leathery while the hind wings are large and membranous, the veins providing strength. The legs are terminated by claws for gripping. The hind leg is particularly powerful; the femur is robust and has several ridges where different surfaces join and the inner ridges bear stridulatory pegs in some species. The posterior edge of the tibia bears a double row of spines and there are a pair of articulated spurs near its lower end. The interior of the thorax houses the muscles that control the limbs.

Crickets, like this great green bush-cricket Tettigonia viridissima, somewhat resemble grasshoppers but have over 20 segments in their antennae and different ovipositors.

The abdomen has eleven segments, the first of which is fused to the thorax and contains the auditory organ and tympanum. Segments two to eight are ring-shaped and joined by flexible membranes. Segments nine to eleven are reduced; segment nine bears a pair of cerci and segments ten and eleven house the reproductive organs. Female grass-hoppers are normally larger than males, with short ovipositors. The name "Caelifera" comes from the Latin and means *chisel-bearing*, referring to the sharp ovipositor.

Those species that make easily heard noises usually do so by rubbing a row of pegs on the hind femurs against the edges of the forewings (stridulation). These sounds are produced mainly by males to attract females, though in some species the females also stridulate.

Grasshoppers are easily confused with the other sub-order of Orthoptera, Ensifera (crickets), but differ in many aspects, such as the number of segments in their anten-nae and structure of the ovipositor, as well as the location of the tympana and modes of sound production. Ensiferans have antennae that can be much longer than the body and have at least 20–24 segments, while caeliferans have fewer segments in their short-er, stouter antennae.

Phylogeny and Evolution

The phylogeny of the Caelifera based on mitochondrial RNA of 32 taxa in six out of seven superfamilies is shown as a cladogram. The Ensifera, Caelifera and all the super-families of grasshoppers except Pamphagoidea appear to be monophyletic.

Fossil grasshoppers at the Royal Ontario Museum

In evolutionary terms, the split between the Caelifera and the Ensifera is no more re-cent than the Permo-Triassic boundary; the earliest insects that are certainly Caelifer-ans are in the extinct families Locustopseidae and Locustavidae from the early Triassic. The group diversified during the Triassic and have remained important plant-eaters from that time to now. The first modern families such as the Eumastacidae, Tetrigidae and Tridactylidae appeared in the Cretaceous, though some insects that might belong to the last two of these groups are found in the early Jurassic. Morphological classifi-cation is difficult because many taxa have converged towards a common habitat type; recent taxonomists have concentrated on the internal genitalia, especially those of the male. This information is not available from fossil specimens, and the palaentological taxonomy is founded principally on the venation of the hindwings.

Diversity and Range

The Caelifera includes some 2,400 valid genera and about 11,000 species. Many undescribed species probably exist, especially in tropical wet forests. The Caelifera have a predominantly tropical distribution with fewer species known from temperate zones, but most of the superfamilies have representatives worldwide. They are almost exclusively herbivorous and are probably the oldest living group of chewing herbivorous insects.

Biology

Diet and Digestion

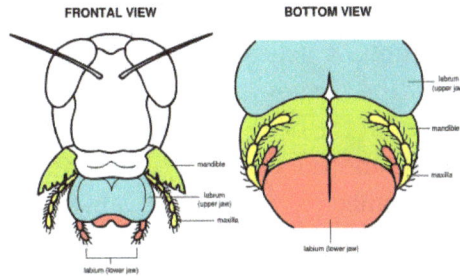

Grasshopper mouth structure

Most grasshoppers are polyphagous, eating vegetation from multiple plant sources, but some are omnivorous and also eat animal tissue and animal faeces. In general their preference is for grasses, including many cereals grown as crops. The mandibles chew the food slightly and salivary glands in the buccal cavity chemically begin to digest the carbohydrates present in it. The food is then passed via the oesophagus to the crop where it is stored temporarily and chemical digestion continues. Next it moves to the gizzard which has muscular walls and tooth-like plates which grind the food. From here, food enters the stomach, where six hepatic caeca add further enzymes and digestion is completed. At the junction between mid and hind-gut, several fine tubes known as malpighian tubules add the excretory products (uric acid, urea and amino acids) to the contents of the gut. Absorption of nutrients takes place in the ileum and any undigested residue is passed on to the colon. Here water is absorbed and the residue becomes solid. After storage in the rectum, the faeces are expelled as small dry pellets.

Sensory Organs

Grasshoppers have a typical insect nervous system, and have an extensive set of external sense organs. On the side of the head are a pair of large compound eyes which give a broad field of vision and can detect movement, shape, colour and distance. There are also three simple eyes (ocelli) on the forehead which can detect light intensity, a pair of antennae containing olfactory (smell) and touch receptors, and mouthparts containing gustatory (taste) receptors. At the front end of the abdomen there is a pair of tympanal organs for sound reception. There are numerous fine hairs covering the whole body that act as mechanoreceptors (touch and wind sensors), and these are most dense on the antennae, the palps (part of the mouth), and on the cerci at the tip of the abdomen. There are special receptors (campaniform sensillae) embedded in the cuticle of the legs that sense pressure and cuticle distortion. There are internal "chordotonal" sense organs specialized to detect position and movement about the joints of the exoskeleton. The receptors convey information to the central nervous system through sensory neurons, and most of these have their cell bodies located in the periphery near the receptor site itself.

Circulation and Respiration

Like other insects, grasshoppers have an open circulatory system and their body cavities are filled with haemolymph. A heart-like structure pumps the fluid to the head from where it percolates past the tissues and organs on its way back to the abdomen. It circulates nutrients throughout the body and carries metabolic wastes to be excreted into the gut. The haemolymph and the circulatory system are not involved in gaseous exchange. Respiration is performed using tracheae, air-filled tubes, which open at the surfaces of the thorax and abdomen through pairs of valved spiracles. Larger insects may need to actively ventilate their bodies by opening some spiracles while others remain closed, using abdominal muscles to expand and contract the body and pump air through the system.

Jumping

A large grasshopper such as a locust can jump about a metre (twenty body lengths) without using its wings; the acceleration peaks at about 20 g. Grasshoppers jump by extending their large back legs and pushing against the substrate (the ground, a twig, a blade of grass or whatever else they are standing on); the reaction force propels them into the air. They jump for several reasons; to escape from a predator, to launch themselves for flight, or simply to move from place to place. For the escape jump in particular there is strong selective pressure to maximize take-off velocity, since this determines the range. This means that the legs must thrust against the ground with both high force and a high velocity of movement. However, a fundamental property of muscle is that it cannot contract with both high force *and* high velocity, which seems like a problem. Grasshoppers overcome this apparent contradiction by using a catapult mechanism to amplify the mechanical power produced by their muscles.

The jump is a three-stage process. First, the grasshopper fully flexes the lower part of the leg (tibia) against the upper part (femur) by activating the flexor tibiae muscle (the back legs of the immature grasshopper in the top photograph are in this preparatory position). Second, there is a period of co-contraction in which force builds up in the large, pennate extensor tibiae muscle, but the tibia is kept flexed by the simultaneous contraction of the flexor tibiae muscle. The extensor muscle is much stronger than the flexor muscle, but the latter is aided by specializations in the joint that give it a large effective mechanical advantage over the former when the tibia is fully flexed. Co-contraction can last for up to half a second, and during this period the extensor muscle shortens and stores elastic strain energy by distorting stiff cuticular structures in the leg. The extensor muscle contraction is quite slow (almost isometric), which allows it to develop high force (up to 14 N in the desert locust), but because it is slow only low power is needed. The third stage of the jump is the trigger relaxation of the flexor muscle, which releases the tibia from the flexed position. The subsequent rapid tibial extension is driven mainly by the relaxation of the elastic structures, rather than by further shortening of the extensor muscle. In this way the stiff cuticle acts like the elastic of a catapult, or the bow of a bow-and-arrow. Energy is put into the store at low power by slow but strong muscle contraction, and retrieved from the store at high power by rapid relaxation of the mechanical elastic structures.

Lifecycle and Reproduction

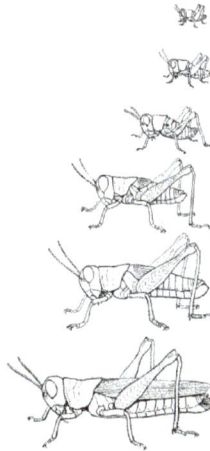

Six stages (instars) of development, from newly hatched nymph to fully winged adult

Grasshoppers lay their eggs in pods in the ground near food plants, generally in the summer. The eggs in the pod are glued together with a froth in some species. After a few weeks of development, the eggs of most species go into diapause, and pass the winter in this state; in a few species the eggs hatch in the same summer they were laid. Diapause is broken by a sufficiently low ground temperature; development resumes as soon as the ground warms above a threshold temperature. The embryos in a pod generally all hatch out within a few minutes of each other. They soon shed their membranes and their exoskeletons harden. These first instar nymphs can then jump away from predators.

Grasshoppers have incomplete metamorphosis: they repeatedly moult (undergo ecdysis), becoming larger and more like an adult, with for instance larger wing-buds, in each instar. The number of instars varies between species. At the final moult, the wings are inflated and become fully functional. The migratory grasshopper, *Melanoplus san-guinipes*, spends about 25–30 days as a nymph depending on sex and temperature, and about 51 days as an adult.

Romalea guttata grasshoppers: female (larger) is laying eggs, with male in attendance.

Males stridulate, rapidly rasping the hind femur against the forewing to create a chur-ring sound, to attract mates. Females select suitable egg-laying sites, such as bare soil or near the roots of food plants according to species. Males often gather around an ovi-positing female; in some species she is mated as soon as she takes her ovipositor out of the ground. After laying the eggs, the female covers the hole with soil and litter.

Predators, Parasites and Pathogens

Cottontop tamarin monkey eating a grasshopper

Grasshoppers have a wide range of predators at different stages of their life-cycle. Eggs are eaten by bee-flies, ground beetles and blister beetles. Hoppers and adults are taken by predators including other insects such as ants, robber flies and sphecid wasps; spi-ders; many birds; and small mammals.

Parasitoids include blowflies, fleshflies, and tachinid flies. External parasites include mites. It has been found that female grasshoppers parasitised by mites produce fewer eggs and thus have fewer offspring. This is probably because the individuals concerned allocate resources in response to the parasitism which are then not available for repro-duction.

Grasshopper with parasitic mites

Spinochordodes tellinii and *Paragordius tricuspidatus* are parasitic worms that infect grasshoppers and alter the behaviour of their hosts. The grasshopper is persuaded to leap into a nearby body of water where it drowns, thus enabling the parasite to continue with the next stage of its life cycle which takes place in water. The grasshopper nematode (*Mermis nigrescens*) is a long slender worm that infests grasshoppers, living in the insect's hemocoel. Adult worms lay eggs on plants and the host gets infected when it eats the foliage.

Locusts killed by the naturally-occurring fungus Metarhizium, an environmentally friendly means of biological control. CSIRO, 2005

Grasshoppers are affected by diseases caused by bacteria, viruses, fungi and protozoa. The bacteria *Serratia marcescens* and *Pseudomonas aeruginosa* have both been implicated in causing disease in grasshoppers, as has the entomopathogenic fungus *Beauveria bassiana*. This widespread fungus has been used to control various pest insects around the world, but although it infects grasshoppers, basking in the sun has the result of raising the insect's temperature above a threshold tolerated by the fungus, and the infection is not lethal. The fungal pathogen *Entomophaga grylli* is able to influence the behaviour of its grasshopper host, causing it to climb to the top of a plant and cling to the stem as it dies. This ensures wide dispersal of the fungal spores liberated from the corpse.

The fungal pathogen *Metarhizium acridum* is found in Africa, Australia and Brazil where it has caused epizootics in grasshoppers. It is being investigated for possible use as a microbial insecticide for locust control. The microsporidian fungus *Nosema locustae*, once considered to be a protozoan, can be lethal to grasshoppers. It has to be consumed by mouth and is the basis for a bait-based commercial microbial pesticide. Various other microsporidians and protozoans are found in the gut.

Anti-predator Defences

Grasshoppers exemplify a range of anti-predator adaptations, enabling them to avoid detection, to escape if detected, and in some cases to avoid being eaten if captured. Grasshoppers are often camouflaged to avoid detection by predators that hunt by sight. Their colouration usually resembles the background, whether green for leafy vegetation, sandy for open areas or grey for rocks. Some species can change their colouration to suit their surroundings.

Gaudy grasshopper, Atractomorpha lata, evades predators with camouflage.

Several species such as the hooded leaf grasshopper *Phyllochoreia ramakrishnai* (Eumastacoidea) are detailed mimics of leaves. Grasshoppers often have deimatic patterns on their wings, giving a sudden flash of bright colours that may startle predators long enough to give time to escape in a combination of jump and flight.

Some species are genuinely aposematic, having both bright warning coloration and sufficient toxicity to dissuade predators. *Dictyophorus productus* (Pyrgomorphidae) is a "heavy, bloated, sluggish insect" that makes no attempt to hide; it has a bright red abdomen. A *Cercopithecus* monkey that ate other grasshoppers refused to eat the species. Another species, the rainbow or painted grasshopper of Arizona, *Dactylotum bicolor* (Acridoidea), has been shown by experiment with a natural predator, the little striped whiptail lizard, to be aposematic.

Lubber grasshopper, Titanacris albipes, has deimatically coloured wings, used to startle predators.

Leaf grasshopper, Phyllochoreia ramakrishnai, mimics a green leaf.

Relationship with Humans

Detail of grasshopper on table in Rachel Ruysch's painting Flowers in a Vase, c. 1685. National Gallery, London

In Art

Grasshoppers are occasionally depicted in artworks, such as the Dutch Golden Age painter Balthasar van der Ast's still life oil painting, *Flowers in a Vase with Shells and Insects*, c. 1630, now in the National Gallery, London, though the insect may be a bush-cricket.

Another orthopteran is found in Rachel Ruysch's still life *Flowers in a Vase*, c. 1685. The seemingly static scene is animated by a "grasshopper on the table that looks about ready to spring", according to the gallery curator Betsy Wieseman, with other invertebrates including a spider, an ant, and two caterpillars.

Symbolism

Sir Thomas Gresham's gilded grasshopper symbol, Lombard Street, London, 1563

Grasshoppers are sometimes used as symbols, as in Sir Thomas Gresham's gilded grasshopper in Lombard Street, London, dating from 1563; the building was for a while the headquarters of the Guardian Royal Exchange, but the company declined to use the symbol for fear of confusion with the locust.

When grasshoppers appear in dreams, these have been interpreted as symbols of "Freedom, independence, spiritual enlightenment, inability to settle down or commit to decision". Locusts are taken literally to mean devastation of crops in the case of farmers; figuratively as "wicked men and women" for non-farmers; and "Extravagance, misfortune, & ephemeral happiness" by "gypsies".

As Food

Hot and sweet crispy grasshoppers, Yogyakarta, Indonesia

Fried grasshoppers from Gunung Kidul, Yogyakarta

In some countries, grasshoppers are used as food. In southern Mexico, grasshoppers, known as *chapulines*, are eaten in a variety of dishes, such as in tortillas with chilli sauce. Grasshoppers are served on skewers in some Chinese food markets, like the Donghuamen Night Market. Fried grasshoppers (*walang goreng*) are eaten in the Gunung Kidul area of Yogjakarta, Java in Indonesia. In the Arab world, grasshoppers are boiled, salted, and sun-dried, and eaten as snacks. In Native America, the Ohlone people burned grassland to herd grasshoppers into pits where they could be collected as food.

It is recorded in the Bible that John the Baptist ate locusts and wild honey while living in the wilderness; attempts have been made to explain the locusts as suitably ascetic vegetarian food such as carob beans.

As Pests

Crop pest: grasshopper eating a maize leaf

Chinese rice grasshopper (*Oxya chinensis*, Borneo

Grasshoppers eat large quantities of foliage both as adults and during their development, and can be serious pests of arid land and prairies. Pasture, grain, forage, vegetable and other crops can be affected. Grasshoppers often bask in the sun, and thrive in warm sunny conditions, so drought stimulates an increase in grasshopper populations. A single season of drought is not normally sufficient to stimulate a massive population increase, but several successive dry seasons can do so, especially if the intervening winters are mild so that large numbers of nymphs survive. Although sunny weather stimulates growth, there needs to be an adequate food supply for the increasing grasshopper population. This means that although precipitation is needed to stimulate plant growth, prolonged periods of cloudy weather will slow nymphal development.

Grasshoppers can best be prevented from becoming pests by manipulating their environment. Shade provided by trees will discourage them and they may be prevented from moving onto developing crops by removing coarse vegetation from fallow land and field margins and discouraging luxurious growth beside ditches and on roadside verges. With increasing numbers of grasshoppers, predator numbers may increase, but this seldom happens sufficiently rapidly to have much effect on populations. Biological control is being investigated, and spores of the protozoan parasite *Nosema locustae* can be used mixed with bait to control grasshoppers, being more effective with immature insects. On a small scale, neem products can be effective as a feeding deterrent and as a disruptor of nymphal development. Insecticides can be used, but adult grasshoppers are difficult to kill, and as they move into fields from surrounding rank growth, crops may soon become reinfested.

Grasshoppers, like the Chinese rice grasshopper, are a pest in rice paddies. Ploughing exposes the eggs on the surface of the field, to be destroyed by sunshine or eaten by natural enemies. Some eggs may be buried too deeply in the soil for hatching to take place.

Locusts

Millions of plague locusts on the move in Australia

Locusts are the swarming phase of certain species of short-horned grasshoppers in the family Acrididae. It has been shown that swarming behaviour is a response to overcrowding. Increased tactile stimulation of the hind legs causes an increase in levels of serotonin. This causes the grasshopper to change colour, feed more and breed faster. The transformation of a solitary individual into a swarming one is induced by several contacts per minute over a short period.

Following this transformation, under suitable conditions dense nomadic bands of flightless nymphs can occur, producing pheromones which attract them to each other. With several generations in a year, the locust population can build up from localised groups into vast accumulations of flying insects known as plagues, devouring all the vegetation they encounter. The largest recorded locust swarm was one of the now-extinct Rocky Mountain locust in 1875, which was 1,800 miles (2,900 km) long and 110 miles (180 km) wide.

An adult desert locust can eat about 2 g (0.1 oz) each day so the billions of insects in a large swarm can be very destructive, stripping all the foliage from plants in an affected area and also consuming stems, flowers, fruits, seeds and bark. Locust plagues can have devastating effects on human populations, causing famines and population upheavals. They are mentioned in both the Koran and the Bible and have been held responsible for cholera epidemics, resulting from the corpses of locusts drowned in the Mediterranean Sea and decomposing on beaches.

The FAO and other organisations monitor locust activity around the world. Timely application of pesticides can prevent nomadic bands of hoppers joining together and proliferating before dense swarms of adults are built up. Besides conventional control using contact insecticides, biological pest control using the entomopathogenic fungus *Metarhizium acridum* which specifically infects grasshoppers has been used with some success.

In Literature

Egyptian hieroglyphs "sn□m"

The Egyptian word for locust or grasshopper was written *sn□m* in the consonantal hieroglyphic writing system. The pharaoh Ramesses II compared the armies of the Hittites to locusts: "They covered the mountains and valleys and were like locusts in their multitude."

Ancient Greek tetradrachm coin from Akragas, 410 BC, with a grasshopper on the right.

One of Aesop's Fables, later retold by La Fontaine, is the tale of *The Ant and the Grasshopper*. The ant works hard all summer, while the grasshopper plays. In winter, the ant is ready but the grasshopper starves. Somerset Maugham's short story "The Ant and the Grasshopper" explores the fable's symbolism via complex framing. The Canadian philosopher Bernard Suits retells the story with the grasshopper as "the exemplification of the life most worth living." Other human weaknesses besides improvidence have become identified with the grasshopper's behaviour. So an unfaithful woman (hopping from man to man) is "a grasshopper" in "Poprygunya", an 1892 short story by Anton Chekhov, and in Jerry Paris's 1969 film *The Grasshopper*.

The 1957 film *Beginning of the End* portrayed giant grasshoppers attacking Chicago. In the 1998 film *A Bug's Life*, the heroes are the members of an ant colony, and the lead villain and his henchmen are grasshoppers.

In aviation

The name "Grasshopper" was used for light aircraft such as the Aeronca L-3 and Piper L-4 used for reconnaissance and other support duties in World War II.

Caddisfly

The caddisflies are an order, Trichoptera, of insects with approximately 7,000 described species. Also called sedge-flies or rail-flies, they are small moth-like insects having two pairs of hairy membranous wings. They are closely related to Lepidoptera (moths and butterflies) which have scales on their wings, and the two orders together form the superorder Amphiesmenoptera.

Caddisflies have aquatic larvae and are found in a wide variety of habitats such as streams, rivers, lakes, ponds, spring seeps and temporary waters (vernal pools). The larvae of many species use silk to make protective cases of gravel, sand, twigs or other debris.

Ecology

Although caddisflies may be found in waterbodies of varying qualities, species-rich caddisfly assemblages are generally thought to indicate clean water. Together with stoneflies and mayflies, caddisflies feature importantly in bioassessment surveys of streams and other water bodies. Caddisfly species can be found in all feeding guilds in stream habitats, with some species being predators, leaf shredders, algal grazers, and collectors of particles from the watercolumn and benthos.

Underwater Architects

Caddisfly larva with portable case of rock fragments

Most caddisfly larvae are underwater architects and use silk, excreted from salivary glands near their mouths, for building. Caddisflies can be divided loosely into three behavioral groups based on their use of silk: net-making caddisflies and case-making cad-disflies, both of which may enlarge their structures throughout their larval lifespan; and free-living caddisflies, which only make such structures prior to pupation. Net-making caddisflies usually live in running water, and their nets, often made amongst aquatic vegetation, serve both as a means to collect algae, detritus, and animal food and as re-treats. Case-making caddisflies may build cases exclusively of silk, but more commonly the silk holds together substrate materials such as small fragments of rock, sand, small pieces of twig or aquatic plants.

Caddisfly larva emerging from case made of plant material

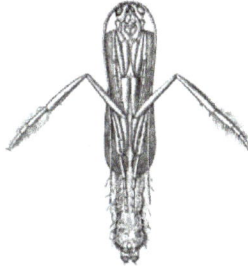

Pupa of Caddisfly in swimming position.
Twice natural size.

Pupa of caddisfly

Caddisfly cases are open at both ends, the larva drawing oxygenated water through the posterior end, over their gills, and out of the wider anterior end. The anterior end is usually wider and it is to this end that they add material as they grow. Their abdomens are soft, but their tougher front ends project from their larval tubes, allowing them to walk while dragging their cases along with them. Caddisfly cases resemble bagworm cases, which are constructed by various terrestrial moth species. Free-living caddisflies do not build retreats or carry portable cases until they are ready to pupate, and their bodies tend to be tougher than caddisflies that build.

Development

Many species of caddisfly larvae enter a stage of inactivity called the pupae stage for weeks or months after they mature but prior to emergence. Their emergence is then triggered by cooling water temperatures in the fall, effectively synchronizing the adult activity to make mate-finding easier. In the Northwestern US, caddisfly larvae within their gravel cases are called 'periwinkles.'

Caddis pupae

Caddisfly pupation occurs much like pupation of Lepidoptera. That is, caddisflies pupate in a cocoon spun from silk. Caddisflies that build the portable cases attach their case to some underwater object, seal the front and back apertures against predation though still allowing water flow, and pupate within it. Once fully developed, most pupal caddisflies cut through their cases with a special pair of mandibles, swim up to the water surface, cast off skin and the now-obsolete gills and mandibles, and emerge as fully formed adults. In a minority of species, the pupae swim to shore—either below the water or across the surface—and crawl out to emerge. Many of them are able to fly immediately after breaking from their pupal skin.

The adult stage of caddisflies, in most cases, is very short-lived, usually only 1–2 weeks, but can sometimes last for 2 months. Most adults are non-feeding and are equipped mainly to mate. Once mated, the female caddisfly will often lay eggs (enclosed in a gelatinous mass) by attaching them above or below the water surface. Eggs hatch in as little as three weeks.

Caddisflies in most temperate areas complete their life cycles in a single year. The general temperate-zone lifecycle pattern is one of larval feeding and growth in autumn, winter, and spring, with adult emergence between late spring and early fall, although the adult activity of a few species peaks in the winter. Larvae are active in very cold water and can frequently be observed feeding under ice. In common with many aquatic insect species, many caddisfly adults emerge synchronously *en masse*. Such emergence patterns ensure that most caddisflies will encounter a member of the opposite sex in a timely fashion. Mass emergences of this nature are called 'hatches' by salmon and trout anglers, and salmonid fish species will frequently 'switch' to whatever species is emerging on a particular day. Anglers take advantage of this behavior by matching their artificial flies to the appropriate fly.

Geological Record

Fossil caddisflies have been found in rocks dating back to the Triassic.

The pupal cases made by caddisflies can be viewed in terms of their fossil records that date back to 250 million years. Old groups of caddisflies are known to have a high level of diversity in their larval cases that are distinctive for each family or genus. For example, a suborder, Integripalpia, have cases that contain plant matter. Whereas the suborder Annulipalpia have cases made from silk and detritus. However, some caddis-flies, specifically among the suborder, Spicipalpia, are free-living with no cases, instead creating a net-like trap with silk.

Artwork

While caddisflies in the wild construct their cases out of twigs, sand, aquatic plants, and rocks, a French artist, Hubert Duprat, makes art by providing wild caddisflies with precious stones and other materials. He collected caddisfly larvae from the wild and put them in climate-controlled tanks. He removes the larvae from their original cases and adds precious and semi-precious items into the tank. The larvae then build new cases out of precious items, creating a unique form of artwork. These works are sold across the world and this technique of making art has been adopted by other artists.

Wasp

A wasp is any insect of the order Hymenoptera and suborder Apocrita that is neither a bee nor an ant. This means that wasps are paraphyletic with respect to bees and ants, and that all three groups are descended from a common ancestor; the Apocrita form a clade.

The most commonly known wasps, such as yellow jackets and hornets, are in the family Vespidae and are eusocial, living together in a nest with an egg-laying queen and non-reproducing workers. Eusociality is favoured by the unusual haplodiploid system of sex determination in Hymenoptera, as it makes sisters exceptionally closely related to each other. However, the majority of wasp species are solitary, with each adult female living and breeding independently. Many of the solitary wasps are parasitoidal, meaning that they raise their young by laying eggs on or in other insects (any life stage from egg to adult). Unlike true parasites, the wasp larvae eventually kill their hosts. Solitary wasps parasitize almost every pest insect, making wasps valuable in horticulture for biological pest control of species such as whitefly in tomatoes and other crops.

Wasps first appeared in the fossil record in the Jurassic, and diversified into many surviving superfamilies by the Cretaceous. They are a successful and diverse group of insects with tens of thousands of described species; wasps have spread to all parts of the world except for the polar regions. The largest social wasp is the Asian giant hornet, at up to 5 centimetres (2.0 in) in length; among the largest solitary wasps is the giant scoliid of Indonesia, *Megascolia procer*. The smallest wasps are solitary chalcid wasps in the family Mymaridae, including the world's smallest known insect, with a body length of only 0.139 mm (0.0055 in), and the smallest known flying insect, only 0.15 mm (0.0059 in) long.

Wasps play many ecological roles. Some are predators, whether to feed themselves or to provision their nests. Many, notably the cuckoo wasps, are kleptoparasites, laying eggs in the nests of other wasps. Wasps have appeared in literature from Classical times, as the eponymous chorus of old men in Aristophanes' 422 BC comedy, *The Wasps*, and in science fiction from H. G. Wells's 1904 novel *The Food of the Gods and How It Came to Earth*, featuring giant wasps with three-inch-long stings. The name "Wasp" has been used for many warships and other military equipment.

Taxonomy and Phylogeny

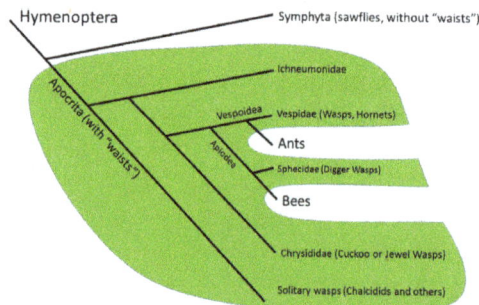

Wasps are paraphyletic, consisting of the clade Apocrita without ants and bees, which are not usually considered to be wasps. The Hymenoptera also contain the somewhat wasplike Symphyta, the sawflies. The familiar common wasps and yellowjackets belong to one family, the Vespidae.

Palaeovespa florissantia, a fossil wasp (Vespinae) from the Eocene rocks of the Florissant fossil beds of Colorado, c. 34 mya

Paraphyletic Grouping

The wasps are a cosmopolitan paraphyletic grouping of hundreds of thousands of species, consisting of the narrow-waisted Apocrita without the ants and bees. The Hymenoptera also contain the somewhat wasplike but unwaisted Symphyta, the sawflies.

The term *wasp* is sometimes used more narrowly for the Vespidae, which includes the common wasp or yellow jacket genera *Vespula* and *Dolichovespula* and the hornets, *Vespa*; or simply for the common wasp and close lookalikes.

Fossils

Male Electrostephanus petiolatus fossil from the Middle Eocene, preserved in Baltic amber

Hymenoptera in the form of Symphyta (Xyelidae) first appeared in the fossil record in the Lower Triassic. Apocrita, wasps in the broad sense, appeared in the Jurassic, and had diversified into many of the extant superfamilies by the Cretaceous; they appear to have evolved from the Symphyta. Fig wasps with modern anatomical features first appeared in the Lower Cretaceous of the Crato Formation in Brazil, some 65 million years before the first fig trees.

The Vespidae include the extinct genus *Palaeovespa*, seven species of which are known from the Eocene rocks of the Florissant fossil beds of Colorado and from fossilised Baltic amber in Europe. Also found in Baltic amber are crown wasps of the genus *Electrostephanus*.

Diversity

Megascolia procer, a giant solitary species from Java in the Scoliidae. This specimen's length is 7.7 cm and its wingspan is 11.5 cm.

Wasps are a diverse group, estimated at over a hundred thousand described species around the world, and a great many more as yet undescribed. For example, there are over 800 species of fig trees, mostly in the tropics, and almost all of these has its own specific chalcid wasp to effect pollination.

Megarhyssa macrurus, a parasitoid. The body of a female is c. 2 inches (51 mm) long, with an ovipositor c. 4 inches (100 mm) long

Many wasp species are parasitoids; the females deposit eggs on or in a host arthropod on which the larvae then feed. Some larvae start off as parasitoids, but convert at a later stage to consuming the plant tissues that their host is feeding on. In other species, the eggs are laid directly into plant tissues and form galls, which protect the developing larvae from predators but not necessarily from other parasitic wasps. In some species, the larvae are predatory themselves; the wasp eggs are deposited in clusters of eggs laid by other insects, and these are then consumed by the developing wasp larvae.

Tarantula hawk wasp dragging an orange-kneed tarantula to her burrow; this species has the most painful sting of any wasp.

The largest social wasp is the Asian giant hornet, at up to 5 centimetres (2.0 in) in length. The tarantula hawk wasp is a similar size and can overpower a spider many times its own weight, and move it to its burrow, with a sting that is excruciatingly painful to humans. The solitary giant scoliid, *Megascolia procer*, with a wingspan of 11.5 cm, has subspecies in Sumatra and Java; it is a parasitoid of the scarabeid Atlas beetle *Chalcosoma atlas*. The female giant ichneumon wasp *Megarhyssa macrurus* is 12.5 centimetres (5 in) long including its very long but slender ovipositor which is used for boring into wood and inserting eggs. The smallest wasps are solitary chalcid wasps in the family Mymaridae, including the world's smallest known insect, *Dicopomorpha echmepterygis* (139 micrometres long) and *Tinkerbella nana* with a body length of only 158 micrometres, the smallest known flying insect.

There are estimated to be 100,000 species of ichneumonoid wasps in the families Braconidae and Ichneumonidae. These are almost exclusively parasitoids, mostly utilising other insects as hosts. Another family, the Pompilidae, is a specialist parasitoid of spiders. Some wasps are even parasitoids of parasitoids; the eggs of *Euceros* are laid beside lepidopteran larvae and the wasp larvae feed temporarily on their haemolymph, but if a parasitoid emerges from the host, the hyperparasites continue their life cycle inside the parasitoid. Parasitoids maintain their extreme diversity through narrow specialism. In Peru, 18 wasp species were found living on 14 fly species in only two species of *Gurania* climbing squash.

Social Wasps

Social wasps constructing a paper nest

Of the dozens of extant wasp families, only the family Vespidae contains any social species, primarily in the subfamilies Vespinae and Polistinae. With their powerful stings and conspicuous warning coloration, often in black and yellow, social wasps are frequent models for Batesian mimicry by non-stinging insects, and are themselves involved in mutually beneficial Müllerian mimicry of other distasteful insects including bees and other wasps. All species of social wasps construct their nests using some form of plant fiber (mostly wood pulp) as the primary material, though this can be supplemented with mud, plant secretions (e.g., resin), and secretions from the wasps themselves; multiple fibrous brood cells are constructed, arranged in a honeycombed pattern, and often surrounded by a larger protective envelope. Wood fibers are gathered from weathered wood, softened by chewing and mixing with saliva. The placement of nests varies from group to group; yellow jackets such as *D. media* and *D. sylvestris* prefer to nest in trees and shrubs; *P. exigua* attaches its nests on the underside of leaves and branches; *Polistes erythrocephalus* chooses sites close to a water source. Other wasps, like *A. multipicta* and *V. germanica,* like to nest in cavities that include holes in the ground, spaces under homes, wall cavities or in lofts. While most species of wasps have nests with multiple combs, some species, such as *Apoica flavissima*, only have one comb. The length of the reproductive cycle depends on latitude; *Polistes erythrocephalus*, for example, has a much longer (up to 3 months longer) cycle in temperate regions.

Solitary Wasps

Potter wasp building mud nest, France. The latest ring of mud is still wet.

The vast majority of wasp species are solitary insects. Having mated, the adult female forages alone and if it builds a nest, does so for the benefit of its own offspring. Some solitary wasps nest in small groups alongside others of their species, but each is involved in caring for its own offspring (except for such actions as stealing other wasp's prey or laying in other wasp's nests). There are some species of solitary wasp that build communal nests, each insect having its own cell and providing food for its own offspring, but these wasps do not adopt the division of labour and the complex behavioural patterns adopted by eusocial species.

Adult solitary wasps spend most of their time in preparing their nests and foraging for food for their young, mostly insects or spiders. Their nesting habits are more diverse than those of social wasps. Many species dig burrows in the ground. Mud daubers and pollen wasps construct mud cells in sheltered places. Potter wasps similarly build vase-like nests from mud, often with multiple cells, attached to the twigs of trees or against walls.

Predatory wasp species normally subdue their prey by stinging it, and then either lay their eggs on it, leaving it in place, or carry it back to their nest where an egg may be laid on the prey item and the nest sealed, or several smaller prey items may be deposited to feed a single developing larva. Apart from providing food for their offspring, no further maternal care is given. Members of the family Chrysididae, the cuckoo wasps, are kleptoparasites and lay their eggs in the nests of unrelated host species.

Biology

Anatomy

Like all insects, wasps have a hard exoskeleton which protects their three main body parts, the head, the mesosoma (including the thorax and the first segment of the abdomen) and the metasoma. There is a narrow waist, the petiole, joining the first and second segments of the abdomen. The two pairs of membranous wings are held together by small hooks and the forewings are larger than the hind ones; in some species, the females have no wings. In females there is usually a rigid ovipositor which may be modified for injecting venom, piercing or sawing. It either extends freely or can be retracted, and may be developed into a stinger for both defence and for paralysing prey.

European hornet, Vespa crabro

In addition to their large compound eyes, wasps have several simple eyes known as ocelli, which are typically arranged in a triangle just forward of the vertex of the head. Wasps possess mandibles adapted for biting and cutting, like those of many other insects, such as grasshoppers, but their other mouthparts are formed into a suctorial proboscis, which enables them to drink nectar.

The larvae of wasps resemble maggots, and are adapted for life in a protected environment; this may be the body of a host organism or a cell in a nest, where the larva either eats the provisions left for it or, in social species, is fed by the adults. Such larvae have soft bodies with no limbs, and have a blind gut (presumably so that they do not foul their cell).

Sand wasp Bembix oculata (Crabronidae) feeding on a fly after paralysing it with its sting

Diet

Adult solitary wasps mainly feed on nectar, but the majority of their time is taken up by foraging for food for their carnivorous young, mostly insects or spiders. Apart from providing food for their larval offspring, no maternal care is given. Some wasp species provide food for the young repeatedly during their development (progressive provisioning). Others, such as potter wasps (Eumeninae) and sand wasps (*Ammophila*, Sphecidae), repeatedly build nests which they stock with a supply of immobilised prey such as one large caterpillar, laying a single egg in or on its body, and then sealing up the entrance (mass provisioning).

Black wasp Sphex pensylvanicus (Sphecidae) feeding on nectar

Predatory and parasitoidal wasps subdue their prey by stinging it. They hunt a wide variety of prey, mainly other insects (including other Hymenoptera), both larvae and adults. The Pompilidae specialize in catching spiders to provision their nests.

Some social wasps are omnivorous, feeding on fallen fruit, nectar, and carrion such as dead insects. Adult male wasps sometimes visit flowers to obtain nectar. Some wasps, such as *Polistes fuscatus*, commonly return to locations where they previously found prey to forage. In many social species, the larvae exude copious amounts of salivary secretions that are avidly consumed by the adults. These include both sugars and amino acids, and may provide essential protein-building nutrients that are otherwise unavailable to the adults (who cannot digest proteins).

Spider wasp (Pompilidae) dragging a jumping spider (Salticidae) to provision a nest

Sex Determination

In wasps, as in other Hymenoptera, sex is determined by a haplodiploid system, which means that females are unusually closely related to their sisters, enabling kin selection to favour the evolution of eusocial behaviour. Females are diploid, meaning that they have 2n chromosomes and develop from fertilized eggs. Males have a haploid (n) number of chromosomes and develop from an unfertilized egg. Wasps store sperm inside their body and control its release for each individual egg as it is laid; if a female wishes to produce a male egg, she simply lays the egg without fertilizing it. Therefore, under most conditions in most species, wasps have complete voluntary control over the sex of their offspring. Experimental infection of *Muscidifurax uniraptor* with the bacterium *Wolbachia* induced thelytokous reproduction and an inability to produce fertile, viable male offspring.

Inbreeding Avoidance

Females of the solitary wasp parasitoid *Venturia canescens* can avoid mating with their brothers through kin recognition. In experimental comparisons, the probability that a female will mate with an unrelated male was about twice as high as the chance of her mating with brothers. Female wasps appear to recognize sibs on the basis of a chemical signature carried or emitted by males. Sib-mating avoidance reduces inbreeding depression that is largely due to the expression of homozygous deleterious recessive mutations.

Ecology

As Pollinators

While the vast majority of wasps play no role in pollination, a few species can effectively transport pollen and are therefore pollinators of several plant species. Since wasps generally do not have a fur-like covering of soft hairs as bees do, pollen does not stick to them well. Pollen wasps in the subfamily Masarinae gather nectar and pollen in a crop inside their bodies, rather than on body hairs like bees, and pollinate flowers of *Penstemon* and the water leaf family, Hydrophyllaceae.

Minute pollinating fig wasps, Pleistodontes: the trees and wasps are mutualistic.

The Agaonidae (fig wasps) are the only pollinators of nearly 1000 species of figs, and thus are crucial to the survival of their host plants. Since the wasps are equally dependent on their fig trees for survival, the relationship is fully mutualistic.

As Parasitoids

The parasitoidal ichneumon wasp Dolichomitus imperator ovipositing through wood, using its immensely long ovipositor to lay its eggs inside hidden larvae, detected by vibration

Many solitary wasps are parasitoids. As adults, these wasps themselves do not take any nutrients from their prey, and those that do feed as adults typically only take nectar from flowers. Parasitoid wasps are extremely diverse in habits, many laying their eggs in inert stages of their host (egg or pupa), sometimes paralysing their prey by injecting it with venom through their ovipositor. They then insert one or more eggs into the host or deposit them upon the outside of the host. The host remains alive until the parasitoid larvae pupate or emerge as adults.

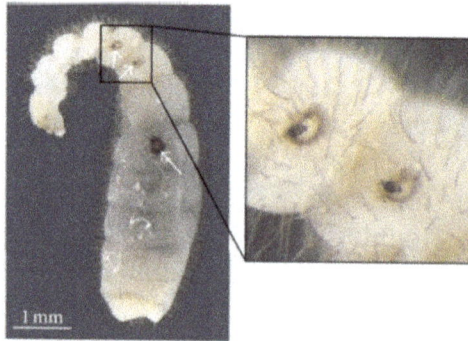

Latina rugosa planidia (arrows, magnified) attached to an ant larva; the Eucharitidae are among the few parasitoids able to overcome the strong defences of ants.

The Ichneumonidae are specialized parasitoids, often of Lepidoptera larvae deeply buried in plant tissues, which may be woody. For this purpose, they have exceptionally long ovipositors; they detect their hosts by smell and vibration. Some of the largest species, including *Rhyssa persuasoria* and *Megarhyssa macrurus*, parasitise horntails, large sawflies whose adult females also have impressively long ovipositors. Some parasitic species have a mutualistic relationship with a polydnavirus that weakens the host's immune system and replicates in the oviduct of the female wasp.

One family of chalcid wasps, the Eucharitidae, has specialized as parasitoids of ants, most species hosted by one genus of ant. Eucharitids are among the few parasitoids that have been able to overcome ants' effective defences against parasitoids.

As Parasites

The Chrysididae, such as this Hedychrum rutilans, are known as cuckoo or jewel wasps for their parasitic behaviour and metallic iridescence.

Many species of wasp, including especially the cuckoo or jewel wasps (Chrysididae), are kleptoparasites, laying their eggs in the nests of other wasp species to exploit their parental care. Most such species attack hosts that provide provisions for their immature stages (such as paralyzed prey items), and they either consume the provisions intended for the host larva, or wait for the host to develop and then consume it before it reaches adulthood. An example of a true brood parasite is the paper wasp *Polistes sulcifer*, which lays its eggs in the nests of other paper wasps (specifically *Polistes dominula*), and whose larvae are then fed directly by the host. Sand wasps *Ammophila* often save time and energy by parasitising the nests of other females of their own species, either kleptoparasitically stealing prey, or as brood parasites, removing the other female's egg from the prey and laying their own in its place. According to Emery's rule, social parasites, especially among insects, tend to parasitise species or genera to which they are closely related. For example, the social wasp *Dolichovespula adulterina* parasitises other members of its genus such as *D. norwegica* and *D. arenaria*.

As Predators

Many wasp lineages, including those in the families Vespidae, Crabronidae, Sphecidae, and Pompilidae, attack and sting prey items that they use as food for their larvae; while Vespidae usually macerate their prey and feed the resulting bits directly to their brood, most predatory wasps paralyze their prey and lay eggs directly upon the bodies, and the wasp larvae consume them. Apart from collecting prey items to provision their young, many wasps are also opportunistic feeders, and will suck the body fluids of their prey. Although vespid mandibles are adapted for chewing and they appear to be feeding on the organism, they are often merely macerating it into submission. The impact of the predation of wasps on economic pests is difficult to establish.

European beewolf Philanthus triangulum provisioning her nest with a honeybee

The roughly 140 species of beewolf (Philanthinae) hunt bees, including honeybees, to provision their nests; the adults feed on nectar and pollen.

As Models for Mimics

The wasp beetle Clytus arietis is a Batesian mimic of wasps.

With their powerful stings and conspicuous warning coloration, social wasps are the models for many species of mimic. Two common cases are Batesian mimicry, where the mimic is harmless and is essentially bluffing, and Müllerian mimicry, where the mimic is also distasteful, and the mimicry can be considered mutual. Batesian mimics of wasps include many species of hoverfly and the wasp beetle. Many species of wasp are involved in Müllerian mimicry, as are many species of bee.

As Prey

While wasp stings deter many potential predators, bee-eaters (in the bird family Meropidae) specialise in eating stinging insects, making aerial sallies from a perch to catch them, and removing the venom from the stinger by repeatedly brushing the prey firmly against a hard object, such as a twig. The honey buzzard attacks the nests of social hymenopterans, eating wasp larvae; it is the only known predator of the dangerous Asian giant hornet or "yak-killer" (*Vespa mandarinia*).

Relationship with Humans

Encarsia formosa, a parasitoid, is sold commercially for biological control of whitefly, an insect pest of tomato and other horticultural crops.

Tomato leaf covered with nymphs of whitefly parasitised by Encarsia formosa

As Pests

Social wasps are considered pests when they become excessively common, or nest close to buildings. People are most often stung in late summer, when wasp colonies stop breeding new workers; the existing workers search for sugary foods and are more likely to come into contact with humans; if people then respond aggressively, the wasps sting. Wasp nests made in or near houses, such as in roof spaces, can present a danger as the wasps may sting if people come close to them. Stings are usually painful rather than dangerous, but in rare cases people may suffer life-threatening anaphylactic shock.

In Horticulture

Some species of parasitic wasp, especially in the Trichogrammatidae, are exploited commercially to provide biological control of insect pests. For example, in Brazil, farmers control sugarcane borers with the parasitic wasp *Trichogramma galloi*. One of the first species to be used was *Encarsia formosa*, a parasitoid of a range of species of whitefly. It entered commercial use in the 1920s in Europe, was overtaken by chemical pesticides in the 1940s, and again received interest from the 1970s. *Encarsia* is used especially in greenhouses to control whitefly pests of tomato and cucumber, and to a lesser extent of aubergine (eggplant), flowers such as marigold, and strawberry. Several species of parasitic wasp are natural predators of aphids and can help to control them. For instance, *Aphidius matricariae* is used to control the peach-potato aphid.

In Sport

Wasps RFC is an English professional rugby union team originally based in London but now playing in Coventry; the name dates from 1867 at a time when names of insects were fashionable for clubs. The club's first kit is black with yellow stripes. The club has an amateur side called Wasps FC.

Among the other clubs bearing the name are a basketball club in Wantirna, Australia, and Alloa Athletic F.C., a football club in Scotland.

In Fashion

Wasp waist, c. 1900, demonstrated by Polaire, a French actress famous for this silhouette

Wasps have been modelled in jewellery since at least the nineteenth century, when diamond and emerald wasp brooches were made in gold and silver settings. A fashion for wasp waisted female silhouettes with sharply cinched waistlines emphasizing the wearer's hips and bust arose repeatedly in the nineteenth and twentieth centuries.

In Literature

The Ancient Greek playwright Aristophanes wrote the comedy play, *The Wasps*, first put on in 422 BC. The "wasps" are the chorus of old jurors.

H. G. Wells made use of giant wasps in his novel *The Food of the Gods and How It Came to Earth* (1904):

It flew, he is convinced, within a yard of him, struck the ground, rose again, came down again perhaps thirty yards away, and rolled over with its body wriggling and its sting stabbing out and back in its last agony. He emptied both barrels into it before he ventured to go near. When he came to measure the thing, he found it was twenty-seven and a half inches across its open wings, and its sting was three inches long. ... The day after, a cyclist riding, feet up, down the hill between Sevenoaks and Tonbridge, very narrowly missed running over a second of these giants that was crawling across the roadway.

Wasp (1957) is a science fiction book by the English writer Eric Frank Russell; it is generally considered Russell's best novel. In Stieg Larsson's book *The Girl Who Played with Fire* (2006) and its film adaptation, Lisbeth Salander has adopted her kickboxing ringname, "The Wasp", as her hacker handle and has a wasp tattoo on her neck, indicating her high status among hackers, unlike her real world situation, and that like a small but painfully stinging wasp, she could be dangerous.

Parasitoidal wasps played an indirect role in the nineteenth-century evolution debate. The Ichneumonidae contributed to Charles Darwin's doubts about the nature and existence of a well-meaning and all-powerful Creator. In an 1860 letter to the American naturalist Asa Gray, Darwin wrote:

I own that I cannot see as plainly as others do, and as I should wish to do, evidence of design and beneficence on all sides of us. There seems to me too much misery in the world. I cannot persuade myself that a beneficent and omnipotent God would have designedly created the Ichneumonidae with the express intention of their feeding within the living bodies of caterpillars, or that a cat should play with mice.

In Military Equipment

With its powerful sting and familiar appearance, the wasp has given its name to many ships, aircraft and military vehicles. Nine ships and one shore establishment of the Royal Navy have been named HMS *Wasp*, the first an 8-gun sloop launched in 1749. Eleven ships of the United States Navy have similarly borne the name USS *Wasp*, the first a merchant schooner acquired by the Continental Navy in 1775. The eighth of these, an aircraft carrier, gained two Second World War battle stars, prompting Winston Churchill to remark "Who said a Wasp couldn't sting twice?" In the Second World War, a German self-propelled howitzer was named Wespe, while the British developed the Wasp flamethrower from the Bren Gun Carrier. In aerospace, the Westland Wasp was a military helicopter developed in England and used by the Royal Navy and other navies; it first flew in 1958. The AeroVironment Wasp III is a Miniature UAV developed for United States Air Force special operations.

HMS *Wasp* (1880), one of nine Royal Navy warships to bear the name

References

- Chapman, A. D. (2006). Numbers of living species in Australia and the World. Canberra: Australian Biological Resources Study. ISBN 978-0-642-56850-2.

- Gullan, P.J.; Cranston, P.S. (2005). The Insects: An Outline of Entomology (3 ed.). Oxford: Blackwell Publishing. ISBN 1-4051-1113-5.

- Merritt, RW; KW Cummins & MB Berg (2007). An Introduction To The Aquatic Insects Of North America. Kendall Hunt Publishing Company. ISBN 0-7575-4128-3.

- Chapman, R.F. (1998). The Insects; Structure and Function (4th ed.). Cambridge, UK: Cambridge University Press. ISBN 0521578906.

- Schowalter, Timothy Duane (2006). Insect ecology: an ecosystem approach (2(illustrated) ed.). Academic Press. p. 572. ISBN 978-0-12-088772-9.

- Evans, Arthur V.; Charles Bellamy (2000). An Inordinate Fondness for Beetles. University of California Press. ISBN 978-0-520-22323-3.

- Holldobler, Wilson (1994). Journey to the ants: a story of scientific exploration. Westminster college McGill Library: Cambridge, Mass.:Belknap Press of Haravard University Press, 1994. pp. 196–199. ISBN 0-674-48525-4.

- Pierce, BA (2006). Genetics: A Conceptual Approach (2nd ed.). New York: W.H. Freeman and Company. p. 87. ISBN 0-7167-8881-0.

- Michels, John (1880). John Michels, ed. Science. 1. American Association for the Advance of Science. 229 Broadway ave., N.Y.: American Association for the Advance of Science. pp. 2090pp. ISBN 1-930775-36-9.

- Hammond, P.M. (1992). "Species inventory". Global Biodiversity, Status of the Earth's Living Resources: a Report (1st ed.). London: Chapman & Hall. pp. 17–39. ISBN 978-0-412-47240-4.

- Foottit, Robert G.; Adler, Peter Holdridge (2009). Insect biodiversity: science and society. John Wiley and Sons. ISBN 1-4051-5142-0.

- Schmidt-Nielsen, Knut (January 15, 1997). "Insect Respiration". Animal Physiology: Adaptation and Environment (5th ed.). Cambridge University Press. p. 55. ISBN 0-521-57098-0.

- Arnett R. H., Jr. & Thomas, M. C. (2001). "Haliplidae". American Beetles, Volume 1. CRC Press, Boca Raton, Florida. pp. 138–143. ISBN 0-8493-1925-0.

- Ramos-Elorduy, Julieta; Menzel, Peter (1998). Creepy crawly cuisine: the gourmet guide to edible insects. Inner Traditions / Bear & Company. p. 5. ISBN 978-0-89281-747-4.

- Wheeler, Alfred George (2001). Biology of the Plant Bugs (Hemiptera: Miridae): Pests, Predators, Opportunists. Cornell University Press. pp. 105–135. ISBN 0-8014-3827-6.

- Alford, David V. (2012). Pests of Ornamental Trees, Shrubs and Flowers: A Color Handbook. Academic Press. p. 12. ISBN 978-0-12-398515-6.

- Dixon, A.F.G. (2012). Aphid Ecology: An optimization approach. Springer Science & Business Media. p. 128. ISBN 978-94-011-5868-8.

- Panizzi, Antônio Ricardo; Parra, José R.P. (2012). Insect Bioecology and Nutrition for Integrated Pest Management. CRC Press. p. 108. ISBN 978-1-4398-3708-5.

- Wheeler, Alfred George (2001). Biology of the Plant Bugs (Hemiptera: Miridae): Pests, Predators, Opportunists. Cornell University Press. pp. 100ff. ISBN 0-8014-3827-6.

- Cranshaw, W. (2013). "11". Bugs Rule!: An Introduction to the World of Insects. Princeton, New Jersey: Princeton University Press. p. 188. ISBN 978-0-691-12495-7.

- Henry, M.S. (2013). Symbiosis: Associations of Invertebrates, Birds, Ruminants, and Other Biota. New York, New York: Elsevier. p. 59. ISBN 978-1-4832-7592-5.

- Grimaldi, D.; Engel, M.S. (2005). Evolution of the insects (1st ed.). Cambridge: Cambridge University Press. p. 237. ISBN 978-0-521-82149-0.

- Bell, W.J.; Roth, L.M.; Nalepa, C.A. (2007). Cockroaches: ecology, behavior, and natural history. Baltimore, Md.: Johns Hopkins University Press. p. 161. ISBN 978-0-8018-8616-4.

- Busvine, J.R. (2013). Insects and Hygiene The biology and control of insect pests of medical and domestic importance (3rd ed.). Boston, MA: Springer US. p. 545. ISBN 978-1-4899-3198-6.

Insect Ecology: An Overview

Insect ecology is concerned with the interaction of insects with the surrounding environment. Insects play a number of roles in the environment, such as pest control, pollination and soil turning and aeration. They are crucial for the biodiversity present on Earth. This section is an overview of the subject matter incorporating all the major aspects of insect ecology.

Insect Ecology

The insect ecology is the scientific study of how insects, individually or as a community, interact with the surrounding environment or ecosystem.

Insects play significant roles in the ecology of the world due to their vast diversity of form, function and life-style; their considerable biomass; and their interaction with plant life, other organisms and the environment. Since they are the major contributor to biodiversity in the majority of habitats, except in the sea, they accordingly play a variety of extremely important ecological roles in the many functions of an eco-system. Taking the case of nutrient recycling; insects contribute to this vital function by degrading or consuming leaf litter, wood, carrion and dung and by dispersal of fungi.

Insects form an important part of the food chain, especially for entomophagous vertebrates such as many mammals, birds, amphibians and reptiles. Insects play an important role in maintaining community structure and composition; in the case of animals by transmission of diseases, predation and parasitism, and in the case of plants, through phytophagy and by plant propagation through pollination and seed dispersal. From an anthropocentric point of view, insects compete with humans; they consume as much as 10% of the food produced by man and infect one in six humans with a pathogen.

Insect Migration

Insect migration is the seasonal movement of insects, particularly those by species of dragonflies, beetles, butterflies and moths. The distance can vary with species and in most cases these movements involve large numbers of individuals. In some cases the individuals that migrate in one direction may not return and the next generation may instead migrate in the opposite direction. This is a significant difference from bird migration.

Monarch butterflies roosting on migration in Texas

Definition

All insects move to some extent. The range of movement can vary from within a few centimeters for some sucking insects and wingless aphids to thousands of kilometres in the case of other insects such as locusts, butterflies and dragonflies. The definition of migration is therefore particularly difficult in the context of insects. A behaviour oriented definition proposed is

Migratory behaviour is persistent and straightened-out movement effected by the animal's own locomotory exertions or by its active embarkation on a vehicle. It depends upon some temporary inhibition of station-keeping responses but promotes their eventual disinhibition and recurrence.

—Kennedy, 1985

This definition disqualifies movements made in the search of resources and which are terminated upon finding of the resource. Migration involves longer distance movement and these movements are not affected by the availability of the resource items. All cases of long distance insect migration concern winged insects.

General Patterns

Migrating butterflies fly within a boundary layer, with a specific upper limit above the ground. The air speeds in this region are typically lower than the flight speed of the insect. These 'boundary-layer' migrants include the larger day-flying insects, and their low-altitude flight is obviously easier to observe than that of most high-altitude wind-borne migrants.

Many migratory species tend to have polymorphic forms, a migratory one and a resident phase. The migratory phases are marked by their well developed and long wings. Such polymorphism is well known in aphids and grasshoppers. In the migratory locusts, there are distinct long and short-winged forms.

The energetic cost of migration has been studied in the context of life-history strategies. It has been suggested that adaptations for migration would be more valuable for insects that live in habitats where resource availability changes seasonally. Others have suggested that species living in isolated islands of suitable habitats are more likely to evolve migratory strategies. The role of migration in gene flow has also been studied in many species. Parasite loads affect migration. Severely infected individuals are weak and have shortened lifespans. Infection creates an effect known as culling whereby migrating animals are less likely to complete the migration. This results in populations with lower parasite loads.

Orientation

Migration is usually marked by well defined destinations which need navigation and orientation. A flying insect needs to make corrections for crosswinds. It has been demonstrated that many migrating insects sense windspeed and direction and make suitable corrections. Day-flying insects primarily make use of the sun for orientation, however this requires that they compensate for the movement of the sun. Endogenous time-compensation mechanisms have been proposed and tested by releasing migrating butterflies that have been captured and kept in darkness to shift their internal clocks and observing changes in the directions chosen by them. Some species appear to make corrections while it has not been demonstrated in others.

Most insects are capable of sensing polarized light and they are able to use the polarization of the sky when the sun is occluded by clouds. The orientation mechanisms of nocturnal moths and other insects that migrate have not been well studied, however magnetic cues have been suggested in short distance fliers.

Recent studies suggest that migratory butterflies may be sensitive to the Earth's magnetic field on the basis of the presence of magnetite particles. In an experiment on the monarch butterfly, it was shown that a magnet changed the direction of initial flight of migrating monarch butterflies. However this result was not a strong demonstration since the directions of the experimental butterflies and the controls did not differ significantly in the direction of flight.

Lepidoptera

Migration of butterflies and moths is particularly well known. The Bogong moth is a native insect of Australia that is known to migrate to cooler climates. The Madagascan sunset moth (*Chrysiridia rhipheus*) has migrations of up to thousands of individuals, occurring between the eastern and western ranges of their host plant, when they become depleted or unsuitable for consumption.

In southern India, mass migrations of many species occur before monsoons. As many as 250 species of butterflies in India are migratory. These include members of the Pieridae and Nymphalidae. The Australian painted lady periodically migrates down the cost of Australia, and occasionally, in periods of strong migration in Australia, migrate to New Zealand.

The monarch butterfly migrates from southern Canada to wintering sites in central Mexico where they spend the winter. In the late winter or early spring, the adult monarchs leave the Transvolcanic mountain range in Mexico to travel north. Mating occurs and the females seek out milkweed to lay their eggs, usually first in northern Mexico and southern Texas. The caterpillars hatch and develop into adults that move north, where more offspring can go as far as Central Canada until the next migratory cycle. The entire annual migration cycle involves five generations.

The painted lady (*Vanessa cardui*) is a butterfly whose annual 15,000 km round trip from Scandinavia and Great Britain to West Africa involves up to six generations.

Orthoptera

Locusts (Schistocerca gregaria) regularly migrate with the seasons.

Short-horned grasshoppers sometime form swarms that will make long flights. These are often irregular and may be related to resource availability and thus not fulfilling some definitions of insect migration. There are however some populations of species such as locusts (*Schistocerca gregaria*) that make regular seasonal movements in parts of Africa; exceptionally, the species migrates very long distances, as in 1988 when swarms flew across the Atlantic ocean.

Odonata

Dragonflies hfsamong the longest distance insect migrants. Many species of *Libellula*, *Sympetrum* and *Pantala* are known for their mass migration. *Pantala flavescens* is thought to make the longest ocean crossings among insects, flying between India and Africa on their migrations. Their movements are often assisted by winds.

Coleoptera

Ladybird beetles such as *Hippodamia convergens*, *Adalia bipunctata* and *Coccinella undecimpunctata* have been noted in large numbers in some places. In some cases, these movements appear to be made in the search for hibernation sites.

Insect Hotel

An insect hotel in a Botanical garden

Insect hotel

Ladybird hotel

Bee hotel in Grimbergen, Belgium

An insect hotel is a manmade structure created from natural materials intended to provide shelter for insects. They can come in a variety of shapes and sizes depending on the specific purpose or specific insect it is catered to. Most consist of several different sections that provide insects with nesting facilities – particularly during winter, offering shelter or refuge for many types of insects.

Purpose

Many insect hotels are used as nest sites by insects including solitary bees and solitary wasps. These insects drag prey to the nest where an egg is deposited. Other insects hotels are specifically designed to allow the insects to hibernate, notable examples include ladybirds (ladybugs) and butterflies. Insects hotels are also popular amongst gardeners and fruit and vegetable growers due to encouraging insect pollination.

Different Hotels for Different Insects

Good materials to construct insect hotels with can include using dry stone walls or old tiles. Drilled holes in the hotel materials also encourage insects to leave larvae to gestate. Therefore, different materials, such as stones and woods are recommended for a wide range and diversity of insect life. Logs and bark, and bound reeds and bamboo are also often used. The various components or sizes of holes to use as entry of an insect hotel attract different species. Ready-made insect hotels are also found at garden centers, and particularly ecological and educational conservational centers and organisations.

Solitary Bees and Wasps

Solitary bees, some wasps and bumblebees do not live within a hive with a queen. There are males and females. A fertilized female makes a nest in wood or stone and borred into the wood in order to construct a nursery.

The most common bee hotel is created from a sawn wooden log or portion of a cut tree trunk in which holes are drilled of different sizes (e.g. 2, 4, 6 and 8 mm), about a few centimeters apart. They attract many bees. The holes have to be tilted slightly so that no rainwater can get in. Stone blocks are also used for this purpose. The holes are drilled quite lengthily into the material but not so far as to create a tunnel to the other side of the wood. Furthermore, the entrances to these access burrows must be smooth enough so that the delicate bodies of the insects are not damaged. Often, with wooden hotels, the exterior is sanded. The best location for a hotel is a warm and sheltered place, such as a southern-facing (in the northern hemisphere) wall or hedge. The first insects are already active towards the end of winter and would be actively seeking for such a place to settle. Other species like to furnish their nests with clay, stone and sand, or in between bricks.

Even a simple bundle of bamboo or reeds, tied or put in an old tin can and hung in a warm place, is very suitable for solitary bees. The bamboo must be cut in a specific way to allow entry for the insects. Often people may add stems of elderberry, rose or blackberry shoots whose marrow can serve as a food source as well.

Butterflies

Butterflies that hibernate like to find sheltered places such as crevices in houses and sheds, or enclosed spaces, such as within bundles of leaves. There are special butterfly enclosures available with vertical slits that take into account the sensitive wings of the animals when they enter them.

Parasitic Insects

By using an insect hotel, parasitic insects are also attracted to make use of the facilities. Cuckoo bees and wasps will lay their eggs within the nests of others in order to provide them with readily available food upon hatching without the parent insect having to provide for them.

Hotels also attract predatory insects which help control unwanted bugs. For example, Earwigs are good to have present in and near fruit trees as they eat the plant lice that may settle on the tree and disturb fruit growth. A terracotta flower pot hung upside-down, filled with bundles of straw or wood wool is an ideal house for earwigs. Ladybirds are easy to cater for by placing many twigs within an open wooden box on its side to provide many small cavities. Ladybirds prefer to hibernate in larger groups so this will encourage many to settle in one specific place. Isopods have their usefulness as scavengers in the garden. These animals like large gaps between stacked bricks and roof tiles to shelter from rain and to hide from predators.

References

- HUFFAKER, CARL B. & GUTIERREZ, A. P. (1999). Ecological Entomology. 2nd Edition (illustrated). John Wiley and Sons. ISBN 0-471-24483-X, ISBN 978-0-471-24483-7.Limited preview on Google Books. Accessed on 09 Jan 2010,
- Arnhold, Tilo (23 October 2012). "Falter mit Migrationshintergrund". Helmholtz-Zentrum für

Umweltforschung UFZ (in German). Retrieved 2016-06-27.

- Tipping, Christopher (May 8, 1995). "Chapter 11: The Longest Migration". Department of Ento-mology & Nematology, University of Florida. Retrieved 8 September 2014.

- Bartel, Rebecca; Oberhauser, Karen; De Roode, Jacob; Atizer, Sonya (February 2011). "Monarch butterfly migration and parasite transmission in eastern North America". Ecology. 92 (2): 342–351. doi:10.1890/10-0489.1. PMID 21618914.

- Buden, Donald W. (2010). "Pantala flavescens (Insecta: Odonata) Rides West Winds into Ngulu Atoll, Micronesia: Evidence of Seasonality and Wind-Assisted Dispersal". Pacific Science. 64 (1): 141–143. doi:10.2984/64.1.141.

Various Equipments used in Entomology

Insects are reduced for certain number of reasons; insect traps are used to directly decrease populations of insects. The mechanisms involved in trapping insects vary, as different insects are attracted to different objects. Some of the equipments used in trapping insects are bottle traps for insects, flight interception traps, malaise traps, moth traps and pitfall traps. The aspects elucidated in this text are of vital importance, and provide a better understanding of entomology.

Insect Trap

Insect traps are used to monitor or directly reduce populations of insects or other arthropods. They typically use food, visual lures, chemical attractants and pheromones as bait and are installed so that they do not injure other animals or humans or result in residues in foods or feeds. Visual lures use light, bright colors and shapes to attract pests. Chemical attractants or pheromones may attract only a specific sex. Insect traps are sometimes used in pest management programs instead of pesticides but are more often used to look at seasonal and distributional patterns of pest occurrence. This information may then be used in other pest management approaches.

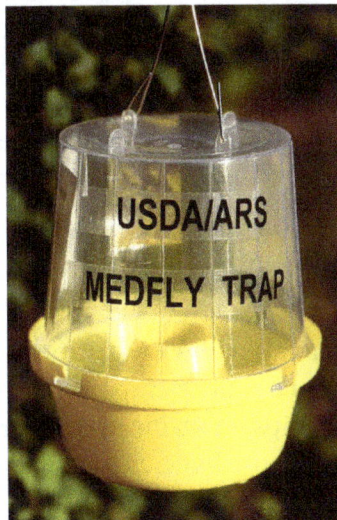

A hanging bucket trap for the Mediterranean fruit fly

The trap mechanism or bait can vary widely. Flies and wasps are attracted by proteins. Mosquitoes and many other insects are attracted by bright colors, carbon dioxide, lactic acid, floral or fruity fragrances, warmth, moisture and pheromones. Synthetic attractants like methyl eugenol are very effective with tephritid flies.

An insect trap mounted onto a pickup truck, for collection of nocturnal species.

Trap Types

Insect traps vary widely in shape, size, and construction, often reflecting the behavior or ecology of the target species. Some common varieties are described below

Light Traps

Light traps, with or without ultraviolet light, attract certain insects. Light sources may include fluorescent lamps, mercury-vapor lamps, black lights, or light-emitting diodes. Designs differ according to the behavior of the insects being studied. Light traps are widely used to survey nocturnal moths. Total species richness and abundance of trapped moths may be influenced by several factors such as night temperature, humidity and lamp type. Grasshoppers and some beetles are attracted to lights at a long range but are repelled by it at short range. Farrow's light trap has a large base so that it captures insects that may otherwise fly away from regular light traps. Light traps can attract flying and terrestrial insects, and lights may be combined with other methods described below.

Adhesive Traps

Sticky traps may be simple flat panels or enclosed structures, often baited, that ensnare insects with an adhesive substance. Sticky traps are widely used in agricultural and indoor pest monitoring. Shelter traps, or artificial cover traps, take advantage of an insect's tendencies to seek shelter in loose bark, crevices, or other sheltered places. Baited shelter traps such "Roach Motels" and similar enclosures often have adhesive material inside to trap insects.

Flying Insect Traps

These traps are designed to catch flying or wind-blown insects.

A sticky insect trap used to monitor pest populations

Flight interception traps or are net-like or transparent structures that impede flying insects and funnel them into collecting. Barrier traps consist of a simple vertical sheet or wall that channels insects down into collection containers. The Malaise trap, a more complex type, is a mesh tent-like trap that captures insects that tend to fly up rather than down when impeded.

Pan traps (also called water pan traps) are simple shallow dishes filled with a soapy water or a preservative and killing agent such as antifreeze. Pan traps are used to monitor aphids and some other small insects.

Bucket traps and bottle traps, often supplemented with a funnel, are inexpensive versions that use a bait or attractant to lure insects into a bucket or bottle filled with soapy water or antifreeze. Many types of moth traps are bucket-type traps. Bottle traps are widely used, often used to sample wasp or pest beetle populations.

Terrestrial Arthropod Traps

Pitfall traps are used for ground-foraging and flightless arthropods such as Carabid beetles and spiders. Pitfall traps consist of a bucket or container buried in soil or other substrate so that its lip is flush with the substrate.

A grain probe is a type of trap used to monitor pests of stored grain, consisting of a long cylindrical tube with multiple holes along its length that can be inserted at various depths within grain.

Soil emergence traps, consisting of an inverted cone or funnel with collecting jar on top, are employed to capture insects with a subterranean pupal stage. Emergence traps have been used to monitor important disease-vectors such as Phlebotomine sandflies.

Aquatic Arthropod Traps

Aquatic interception traps typically involve mesh funnels that or conical structures that guide insects into a jar or bottle for collecting.

Aquatic emergence traps are cage-like or tent-like structures used to capture aquatic insects such as chironomids, caddisflies, mosquitoes, and odonates upon their transition from aquatic nymphs to terrestrial adults. Aquatic emergence traps may be free floating on the water's surface, submerged, or attached to a post near shore.

Bottle Trap for Insects

Bottle trap is a name used for several different objects. Among these are a piece of material used in bathroom plumbing, as well as various traps that are made out of discarded bottles and which are used to trap animals as different as beetles, mice, fish and octopuses. This article is about the use of modified bottles to trap flying insects.

In this context, a bottle trap is a type of baited arboreal insect trap for collecting either prized or harmful frugivorous beetles, especially flower beetles, leaf chafers and longhorn beetles as well as wasps and other unwanted flying insects.

Structure

A bottle trap is an insect trap made out of a plastic bottle. Most collectors use bottles of 1.5 or 2 liters to make these traps but smaller bottles are sometimes used as well. There are basically two types:

- Funnel type. These bottle traps are made by cutting off the neck of the bottle as well as the complete tapering part of the top. The neck and cap are discarded. For catching wasps only the cap is removed, while leaving the neck in place. The tapering part is placed upside down on top of the rest of the bottle, thereby effectively forming a funnel. This funnel is then fixed to the bottle by piercing both bottle and funnel at two opposing sides. A wire fitted through these holes ensures the funnel solidly fits on the bottle, while the trap can easily be opened when required. After putting the bait in the bottle the trap is placed at the desired location.

    o   Advantages:

        ▪   Insects can't escape from this type of trap, since they fly up along the side of the bottle, not finding the exit, which is in the middle.

        ▪   Bats and large moths can't enter the trap, since they are too large to fit through the funnel.

        ▪   This type can, unlike the other kind, also be used to collect troublesome wasps

    o   Disadvantages:

- Not only insects but also rain will funnel into the trap. This trap is therefore normally only used in dry seasons.

- This construction requires a bit more work than the side door type.

- Side-door type. A side-door bottle trap consists of a bottle with cap of which the higher end of one upright side is cut open. A simple rectangular shape is cut out, taking care that it stays attached to the bottle on its upside. This plastic flap is then bend upward, effectively forming a rain shield over the entrance. After adding some bait the trap is put in its place.

  o Advantages:

     - Because of its opening with rain shield very little rain enters the trap, making it effective in wet seasons too.

     - Construction is very simple and requires no additional materials.

  o Disadvantages:

     - The wider opening allows for small bats and large moths to enter. These may die in the trap and pollute it, as well as forming, with their wings, a bridge to the exit.

     - Captured beetles may escape again since they may simply fly upward along the side of the bottle.

Bait

Many different types of bait are used. Since this kind of trap is mainly used for beetles that are attracted to (over)ripe fruits, baits with a certain amount of alcohol are usually very effective. Types of bait which are commonly used are:

- For beetles:

  o Banana with or without beer or rum (and sometimes with added sugar).

Bread soaked with beer will attract roaches.

  o A mixture of red wine, vinegar and sugar.

- For wasps (funnel type bottle trap only):

  o Syrup, soft drink or sugar water.

Other fruits are sometimes used as well, but banana is most often used since it is widely available, normally inexpensive and contains sufficient sugar to start a fermentation process by itself. The different ingredients are usually kept apart and mixed in the trap itself, but some collectors prefer to mix their bait before going into the field.

Placement

Bottle traps (like all traps) yield best in places where more of the desired insects are to be expected. For beetles, in general this means high up in trees, especially flowering or fruiting trees. Other places in which traps are often placed with good results include forest borders. Traps placed inside forests usually yield smaller numbers of beetles, but also different species. Traps for luring wasps are usually set up a short distance (several meters) from the place where they are bothersome.

There are various methods used for placing bottle traps:

- After a hole is made in the top of the trap, a branch is bent down and its top fitted through the hole.

- A thin cord is attached to the trap, after which the cord is thrown over a high branch and the trap pulled up, the rope then being fixed to some lower branch.

- After a wire hook is fixed to it, the trap is either manually or with the aid of a long stick hung over a branch.

- The trap is placed on the ground.

The first three methods are used most often for collecting beetles, while the latter two are more commonly in use for catching wasps.

Bycatch

Next to the desired beetles, many other insects may find the bait attractive. Sap beetles (a group of small fruit-eating beetles), moths like the large white witch moth, various butterflies, cockroaches, flies, stingless bees, wasps and even small fruit eating bats may enter the bottle traps as bycatch while the collector aims for beetles. Such unwanted animals in the trap may cause the collector several problems:

- These unwanted insects pollute the trap. Because they cannot escape the trap, they will, like the beetles, eventually die in it. Fragile insects such as moths and butterflies will live only for a short time in these traps, and after they die their bodies will quickly start rotting, and their intestines as well as the scales on their wings will pollute the bait. Although this will not necessarily alter the attractiveness of the bait to other insects, the remains of the unwanted specimen will make it more difficult for the collector to check the bait for desired beetles. Unwanted animals which are still alive usually need some time to get over the effects of alcohol, after which they will fly away.

- They may create an escape route for trapped beetles. Especially the very long wings of the neotropical white witch moth may form an effective bridge from the bait to the opening of the trap.

- Protected species such as bats may enter the trap and not be able to leave it again. When they are still alive, bats are often able to fly away directly after being flushed with clean water.

Flight Interception Trap

A hanging trap in a forest

The Flight Interception Trap or FIT is a widely used trapping system for flying insects. It is especially well-suited for collecting beetles, since these animals usually drop themselves after flying into an object, rather than flying upward (in which case a Malaise trap is a better option). Flight Interception Traps are mainly used to collect flying species which are not likely to be attracted to bait or light.

Construction

The basis of any Flight Interception Trap consists of an upright placed see-through barrier under which one or more small basins are placed. The barrier may consist of such materials as plastic mesh, a transparent plastic sheet or even Plexiglas, although the latter does not work well for day-active insects since it is visible to them due to its specific refraction. The basins are filled with a preserving fluid such as ethanol (which should be mixed with something bad-tasting (like denatonium) to prevent wild animals drinking it), propylene glycol, salt-saturated water or even plain water. The best preservative to keep internal organs in good condition is FAACC, a solution of formaldehyde. A small amount of detergent is added to break the surface tension, causing the insects to sink. Yellow (a colour which attracts many insects) pans with soapy water may be used alone. The water itself can be an attractant in dry environments .

Location

Depending on either the desired information (for research) or desired species (for trade) the construction can be put in open land or in the forest. It is important to place the barrier in a straight angle with the most likely flying route for insects (e.g. blocking a forest corridor), such as to maximize results.

Checking

The basins can be checked daily (when it is e.g. important to check the activity of the desired insects under different weather conditions), weekly or even less often. Maximum time between two checks depends on the used preservatives, since not all preservatives are equally suited for preserving insects for a longer time.

Cover

To prevent the basins from filling up with litter, most researchers place some kind of roof over the trap. This keeps leaves from falling in while it also keeps the rain out (which could otherwise dilute the preservative or cause an overflow).

Killing Jar

A killing jar is a device used by entomologists to kill captured insects quickly and with minimum damage. The jar, typically glass, must be hermetically sealable and one design has a thin layer of hardened plaster of paris on the bottom to absorb the killing agent. The killing agent will then slowly evaporate, allowing the jar to be used many times before needing to refresh the jar. The absorbent plaster of paris layer also helps prevent the agent sticking to and damaging insects. Crumpled paper tissue is also placed in the jar for the same reason. A second method utilises a wad of cotton or other absorbent material placed in the bottom of the jar. Liquid killing agent is then added until the absorbent material is nearly saturated. A piece of stiff paper or cardboard cut to fit the inside of the jar tightly is then pressed in.

The most common killing agents are ether, chloroform and ethyl acetate. Ethyl acetate has many advantages and is very widely used. Its fumes are less toxic to humans than those of the other agents and specimens will remain limp if they are left in an ethyl acetate killing jar for several days and the ethyl acetate is not allowed to entirely evaporate from the specimens. A disadvantage is that although the insects are quickly stunned by ethyl acetate it kills them slowly and specimens may revive if removed from the killing jar too soon. Potassium cyanide or other cyanide compounds including calcium cyanide are also used, but only by experts due to its extreme toxicity. It also has the disadvantages of making the specimens brittle when left in the jar for several hours and may also cause some discoloration of colored specimens. It does kill rapidly and the cyanide charge will last a long time. A few drops of acetic acid will increase the cyanide gas production. If the jar is not used for long periods it may dry out and produce little gas, therefore a few drops of water will also help get the process going again. The potassium cyanide slowly decomposes, releasing hydrogen cyanide.

Killing jars are only used on hard-bodied insects. Soft-bodied insects, such as the larval stage, are generally fixed in ethanol at 70-80% concentration.

Malaise Trap

A malaise trap

A Malaise trap is a large, tent-like structure used for trapping flying insects, particularly Hymenoptera and Diptera. The trap is made of a material such as terylene netting and can be various colours. Insects fly into the tent wall and are funnelled into a collecting vessel attached to highest point. It was invented by René Malaise in 1934.

Structure

There are many versions of the Malaise trap, but the basic structure consists of a tent with a large opening at the bottom for insects to fly into and a tall central wall that directs the flying insects upwards to a cylinder containing a killing agent. The chemicals vary according to purpose and access. Conventionally, cyanide was used inside the jar with an absorbent material. However, due to restrictions, many people use ethanol. Ethanol will damage some flying insects like Lepidopterans, but most people use the malaise trap primarily for Hymenopterans and Dipterans. In addition, the ethanol will keep the specimens preserved for a longer period of time. Other dry killing agents include no-pest strips (vapona) and ethyl acetate and need to be checked more regularly.

Design Details

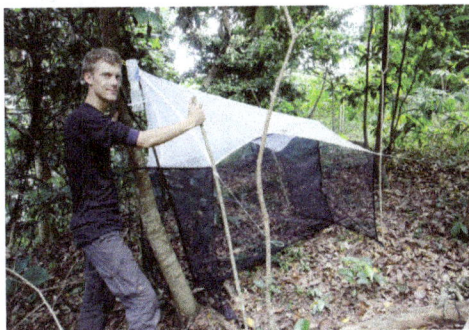

Setting up a malaise trap in Udzungwa Mountains National Park.

Cylinder

When choosing a Malaise trap design, it is important to consider the types of insects you want to catch. The opening to the cylinder is of key importance. Typically, the opening is around 12–15 mm (0.47–0.59 in), and can vary according to the size of insect you are trying to catch. If using a dry agent, a smaller hole will result in a faster death, limiting the amount of damage a newly caught insect will inflict on older, fragile specimens. In ethanol, this is less of a concern. Larger holes also allow in more butterflies, moths, and dragonflies potentially.

Location

Placement of the trap is very important. The trap should be positioned to maximize the number of flying insects that pass through the opening. This is determined by the natural features of the site. One should evaluate topography, vegetation, wind, and water. For example, if there is a wide corridor in a forest such as a trail, the trap should be oriented with its opening to the corridor. Also, places where vegetation is growing high around the opening will limit the number of flying insects that enter the trap. Other ideal places may be above small streams or on edges of forests.

A well placed trap in ideal seasonal conditions can catch over 1,000 insects a day. Even in less ideal conditions, like rain, the trap is still effective.

Other Uses

The Malaise trap can also function as a light trap. If a lamp is placed at the end opposite of the opening, the light will attract insects into the trap. For those who want to know what insects they are catching in the day versus the night, the specimens should be collected and removed at dawn and dusk. For others, specimens should be removed from the trap at least once a week if using ethanol, or more often if using a dry killing agent.

The design of the trap catches insects that naturally fly upwards when they hit a barrier. However, some insects drop when met with a barrier. An addition of a pan with ethanol at the bottom of the main wall will catch specimens like beetles that fall before reaching the top. A trap without the netting on top but with just a preservative-filled basin under the barrier is commonly named a Flight Interception Trap or FIT.

Moth Trap

Moth traps are devices used by entomologists to capture moths. Most use a light source (light trap).

All moth traps follow the same basic design - consisting of a mercury vapour or actinic light to attract the moths and a box in which the moths can accumulate and be examined later. The moths fly towards the light and spiral down towards the source of the light and are deflected into the box. Besides moths, several other insects will also come to light, such as scarab beetles, Ichneumonid wasps, stink bugs, stick insects, diving beetles, and water boatmen. Occasionally diurnal species such as dragonflies, yellowjacket wasps, and hover flies will also visit.

A commercially produced Robinson trap.

A simple moth trap

The reason insects and especially particular families of insect (e.g. moths), are attracted to light is uncertain . The most accepted theory is that moth migrate using the moon and stars as navigational aids and that the placement of a closer than the moon light causes subtended angles of light at the insects eye to alter so rapidly that it has to fly in a spiral to reduce the angular change -- this resulting in the insect flying into the light. Yet the reason some diurnal insects visit is entirely unknown.

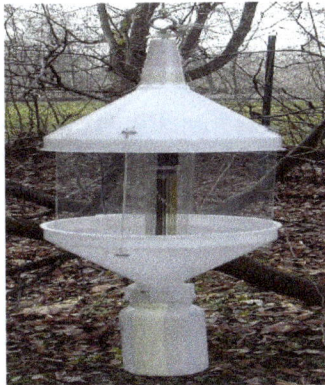

A more complex moth trap

Some moths, notably Sesiidae are monitored or collected using pheromone traps.

Pheromone Trap

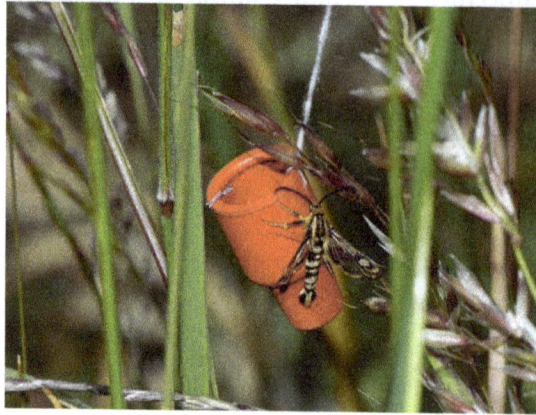

Chamaesphecia empiformis (Sesiidae) on a red rubber septa pheromone lure

A pheromone trap is a type of insect trap that uses pheromones to lure insects. Sex pheromones and aggregating pheromones are the most common types used. A pheromone-impregnated lure, as the red rubber septa in the picture, is encased in a conventional trap such as a Delta trap, water-pan trap, or funnel trap.

Sensitivity

Pheromone traps are very sensitive, meaning they attract insects present at very low densities. They are often used to detect presence of exotic pests, or for sampling, monitoring, or to determine the first appearance of a pest in an area. They can be used for legal control, and are used to monitor the success of the Boll Weevil Eradication Program and the spread of the gypsy moth. The fact that pheromone traps are highly species-specific can also be an advantage, and they tend to be inexpensive and easy to implement.

However, it is impractical in most cases to completely remove or "trap out" pests using a pheromone trap. Some pheromone-based pest control methods have been successful, usually those designed to protect enclosed areas such as households or storage facilities. There has also been some success in mating disruption. In one form of mating disruption, males are attracted to a powder containing female attractant pheromones. The pheromones stick to the males' bodies, and when they fly off, the pheromones make them attractive to other males. It is hoped that if enough males chase other males instead of females, egg-laying will be severely impeded.

Some difficulties surrounding pheromone traps include sensitivity to bad weather, their ability to attract pests from neighboring areas, and the fact that they generally only attract adults although it is the juveniles in many species that are pests. They are also generally limited to one sex.

Pitfall Trap

A Barber pitfall trap

A pitfall trap is a trapping pit for small animals, such as insects, amphibians and reptiles. Pitfall traps are mainly used for ecology studies and ecologic pest control. Animals that enter a pitfall trap are unable to escape. This is a form of passive collection, as opposed to active collection where the collector catches each animal (by hand or with a device such as a butterfly net). Active collection may be difficult or time consuming, especially in habitats where it is hard to see the animals such as thick grass.

Structure and Composition

Pitfall traps come in a variety of sizes and designs. They come in 2 main forms: dry and wet pitfall traps. Dry pitfall traps consist of a container (tin, jar or drum) buried in the ground with its rim at surface level used to trap mobile animals that fall into it. Wet pitfall traps are basically the same, but contain a solution designed to kill and preserve the trapped animals. The fluids that can be used in these traps include formalin (10% formaldehyde), methylated spirits, alcohol, ethylene glycol, trisodium phosphate, picric acid or even (with daily checked traps) plain water. A little detergent is usually added to break the surface tension of the liquid to promote quick drowning. The opening is usually covered by a sloped stone or lid or some other object. This is done to reduce the amount of rain and debris entering the trap, and to prevent animals in dry traps from drowning (when it rains) or overheating (during the day) as well as to keep out predators.

One or more fence-lines of some sort may be added to channel targets into the trap.

Traps may also be baited. Lures or baits of varying specificity can be used to increase the capture rate of a certain target species or group by placing them in, above or near the trap. Examples of baits include meat, dung, fruit and pheromones.

Uses

Pitfall traps can be used for various purposes:

- During the mating seasons of toads, frogs and salamanders in temperate climates, these animals often have to cross busy roads on their way from wintering grounds to breeding ponds. To prevent them from being killed, volunteers may place low fences along roads which the animals have to cross. On short distances, dry pitfall traps are then placed along the fences to collect the animals, which subsequently are manually transferred to the other side of the road, thus preventing massive roadkill.

- Collectors and researchers of various ground-dwelling arthropod species may use pitfall traps to collect the animals they are interested in. This can be done without bait (for example ground beetles and spiders) or with bait (for example dung beetles).

- When used in series, these traps may also be used to estimate species richness (number of species present) and abundances (number of individuals), and this combined information may be used to calculate biodiversity indices (e.g. the Shannon index).

Problems

There are inevitably biases in pitfall sampling when it comes to comparison of different groups of animals and different habitats in which the trapping occurs. An animal's trappability depends on the structure of its habitat (e.g. density of vegetation, type of substrate). Gullan and Cranston (2005) recommend measuring and controlling for such variations. Intrinsic properties of the animal itself also affect its trappability: some taxa are more active than others (e.g. higher physiological activity or ranging over a wider area), more likely to avoid the trap, less likely to be found on the ground (e.g. tree-dwelling species that occasionally move across the terrain), or too large to be trapped (or large enough to escape if trapped). Trappability can also be affected by conditions such as temperature or rain, which may alter the animal's behaviour. The capture rate is therefore proportional not only to how abundant a given type of animal is (which is often the factor of interest), but how easily they are trapped. Comparisons between different groups must therefore take into account variation in habitat structure and complexity, changes in ecological conditions over time and the innate differences in species.

Tullgren Funnel

A Tullgren funnel, also known as Berlese funnel or Berlese trap, is an apparatus used to extract living organisms, particularly arthropods, from samples of soil. The Tullgren funnel works by creating a temperature gradient over the sample such that mobile organisms will move away from the higher temperatures and fall into a collecting vessel, where they perish and are preserved for examination. The illustration shows how it works: a funnel (E) contains the soil or litter (D), and a heat source (F) such as an electric lamp (G) heats the litter. Animals escaping from the desiccation of the litter descend through a filter (C) into a preservative liquid (A) in a receptacle (B). This illustration is merely a schematic, since usually the soil sample will not be crumbled and poured into the funnel (this would inevitably lead to a high amount of soil particles in the preservation fluid requiring laborious work to sort out the soil organisms). In fact, the soil sample is placed on a mesh sieve that will allow the soil animals to pass but should retain most of the soil particles.

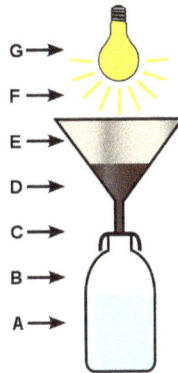

The Berlese funnel is used to extract organisms from soil.

This type of extraction is commonly referred to as Berlese funnel or Tullgren funnel. Antonio Berlese described this method of dynamic sampling in 1905 with a hot water jacket as heat source. In 1918 Albert Tullgren described a modification, where the heating came from above by an electric bulb and the heat gradient was increased by an iron sheet drum around the soil sample. Today's extraction funnels of this type usually combine elements from both publications and thus should be referred to as Berlese-Tullgren funnel.

Electrical Penetration Graph

The Common Brown Leafhopper (Orosius orientalis) connected to the EPG electrode

The electrical penetration graph or EPG is a system used by biologists to study the interaction of insects such as aphids, thrips, and leafhoppers with plants. Therefore, it can also be used to study the basis of plant virus transmission, host plant selection by insects and the way in which insects can find and feed from the phloem of the plant. It is a simple system consisting of a partial circuit which is only completed when a species such as aphids, which are the most abundantly studied, inserts its stylet into the plant in order to *probe* the plant as a suitable host for feeding. The completed circuit is displayed visually as a graph with different waveforms indicating either different insect activities such as saliva excretion or the ingestion of cellular contents or indicating which tissue type has been penetrated (i.e. phloem, xylem or mesophyll). So far, around ten different graphical waveforms are known, correlating with different insect/plant interaction events.

The Circuit

The circuit connects to the insect via a 20 μm gold wire and to the plant via a copper electrode placed in the soil. The circuit also passes through, normally, a one gigaohm resistor and a 50x amplifier before the results are stored digitally and interpreted by a computer to calculate the final graph.

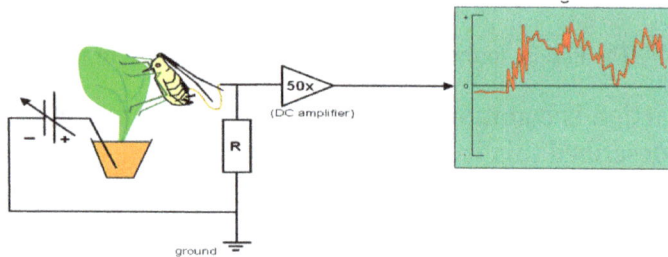

References

- Nancy D. Epsky, Wendell L. Morrill, Richard W. Mankin (2008). "Traps for Capturing Insects". In Capinera, John L. (Ed.). Encyclopedia of Entomology (PDF). Dordrecht: Springer. pp. 3887–3901. ISBN 1402062427.

- Price, B.; Baker, E. (2016). "NightLife: A cheap, robust, LED based light trap for collecting aquatic insects in remote areas .". Biodiversity Data Journal. 4 (e7648): 1–18. doi:10.3897/BDJ.4.e7648.

- Jonason, Franzén and Ranius (2014) Surveying Moths Using Light Traps: Effects of Weather and Time of Year. Plos One, 9, e92453.

- Ellis, M. V.; Bedward, M. (2014). "A simulation study to quantify drift fence configuration and spacing effects when sampling mobile animals". Ecosphere. 5 (5): art55. doi:10.1890/ES14-00078.1.

- Giordanengo P. (2014). EPG-Calc: a PHP-based script to calculate electrical penetration graph (EPG) parameters. Arthropod-Plant Interactions, 8(2):163-169.

- Nora Dunn. "Pesky Pests: Easy Homemade Mosquito and Insect Traps and Repellent". Wise Bread. Retrieved 2013-11-14.

- Ellis, M.V. (2013). "Impacts of pit size, drift fence material and fence configuration on capture rates of small reptiles and mammals in the New South Wales rangelands". Australian Zoologist. 36: 404–412. doi:10.7882/AZ.2013.005.

- Sauge M.H.; Lambert P.; Pascal T. (2012). Co-localisation of host plant resistance QTLs affecting the performance and feeding behaviour of the aphid Myzus persicae in the peach tree. Heredity, 108 (3), 292-301.

- "Flight Interception Traps - Collecting Methods - Mississippi Entomological Museum Home". Mississippientomologicalmuseum.org.msstate.edu. 2011-02-23. Retrieved 2013-11-14.

Applications of Entomology

Methods used for controlling pests are known as biological control. Beekeeping, sericulture and biomimetics are some of the applications of entomology that have been elucidated in this text. The diverse applications of entomology in the current scenario have been thoroughly discussed in the following chapter.

Biological Pest Control

Syrphus hoverfly larva feeding on aphids

Biological control is a method of controlling pests such as insects, mites, weeds and plant diseases using other organisms. It relies on predation, parasitism, herbivory, or other natural mechanisms, but typically also involves an active human management role. It can be an important component of integrated pest management (IPM) programs.

There are three basic types of biological pest control strategies: importation (sometimes called classical biological control), in which a natural enemy of a pest is introduced in the hope of achieving control; augmentation, in which locally-occurring natural enemies are bred and released to improve control; and conservation, in which measures are taken to increase natural enemies, such as by planting nectar-producing crop plants in the borders of rice fields.

Parasitic wasp Cotesia congregata on tobacco hornworm Manduca sexta

Natural enemies of insect pests, also known as biological control agents, include predators, parasitoids, and pathogens. Biological control agents of plant diseases are most often referred to as antagonists. Biological control agents of weeds include seed predators, herbivores and plant pathogens.

Biological control can have side-effects on biodiversity through predation, parasitism, pathogenicity, competition, or other attacks on non-target species, especially when a species is introduced without thorough understanding of the possible consequences.

History of Biological Control

The term "biological control" was first used by Harry Scott Smith at the 1919 meeting of the Pacific Slope Branch of the American Association of Economic Entomologists, at the Mission Inn in downtown Riverside, California; and later defined by P. DeBach and K. S. Hagen in 1964. However, the practice has previously been used for centuries. The first report of the use of an insect species to control an insect pest comes from "Nan Fang Cao Mu Zhuang" (南方草木 ・ *Plants of the Southern Regions*) (ca. 304 AD), which is attributed to Western Jin dynasty botanist *Ji Han* (・ 含, 263-307), in which it is mentioned that *"Jiaozhi people sell ants and their nests attached to twigs looking like thin cotton envelopes, the reddish-yellow ant being larger than normal. Without such ants, southern citrus fruits will be severely insect-damaged"*. The ants used are known as *huang gan* (*huang* = yellow, *gan* = citrus) ants (*Oecophylla smaragdina*). This practice has later also been reported by *Ling Biao Lu Yi* (late Tang Dynasty or Early Five Dynasties), in "Ji Le Pian" by *Zhuang Jisu* (Southern Song Dynasty), in the "Book of Tree Planting" by *Yu Zhen Mu* (Ming Dynasty), in the book "Guangdong Xing Yu" (17th century), "Lingnan" by *Wu Zhen Fang* (Qing Dynasty), in "Nanyue Miscellanies" by *Li Diao Yuan*, and others, which shows that this practice has obviously perdured for a very long time.

The use of Biological control techniques as we know them today started to emerge in the 1870s. During this decade, in the USA, The Missouri State Entomologist C. V. Riley and the Illinois State Entomologist W. LeBaron began within-state redistribution of parasitoids to control crop pests. The first international shipment of an insect as biological control agent was made by Charles V. Riley in 1873, shipping to France the predatory mites *Tyroglyphus phylloxera* to help fight the grapevine phylloxera (*Daktulosphaira vitifoliae*) that was destroying grapevines in France. The United States Department of Agriculture (USDA) initiated research in classical biological control following the establishment of the Division of Entomology in 1881, with C. V. Riley as Chief. The first importation of a parasitoid into the United States was this of *Cotesia glomerata* in 1883-1884, imported from Europe to control the imported cabbage white butterfly, *Pieris rapae*. In 1888-1889 the vedalia beetle, *Rodolia cardinalis*, which is a ladybug, was imported from Australia and introduced into California to control the cottony cushion scale, *Icerya purchasi*, which had become a major problem for the newly developed citrus industry in California, and by the end of 1889 the cottony cushion scale population had already declined. This great success led to further introductions of beneficial insects into the USA.

In 1905 the USDA initiated its first large-scale biological control program, sending entomologists to Europe and Japan to look for natural enemies of the gypsy moth, *Lymantria dispar dispar*, and brown-tail moth, *Euproctis chrysorrhoea*, invasive pests of trees and shrubs. As a result, nine species of parasitoid of gypsy moth, seven of brown-tail moth, and two predators for both moths became established in the USA. Although the gypsy moth was not fully controlled by these natural enemies, the frequency, duration, and severity of its outbreaks were reduced and the program was regarded as successful. This program also led to the development of many concepts, principles, and procedures for the implementation of biological control programs.

The first reported case of a classical biological control attempt in Canada involves the hymenopteran parasitoid *Trichogramma minutum*. Individuals were caught in New York State and released in Ontario gardens in 1882 by William Saunders, trained chemist and first Director of the Dominion Experimental Farms, for controlling the imported currantworm *Nematus ribesii*. Between 1884 and 1908, the first Dominion Entomologist, James Fletcher, continued introductions of other parasitoids and pathogens for the control of pests in Canada.

Types of Biological Pest Control

There are three basic biological pest control strategies: importation (classical biological control), augmentation and conservation.

Importation

Importation or classical biological control involves the introduction of a pest's natural enemies to a new locale where they do not occur naturally. Early instances were often unofficial and not based on research, and some introduced species became serious pests themselves.

Rodolia cardinalis, the vedalia beetle, was imported to Australia in the 19th century, successfully controlling cottony cushion scale.

To be most effective at controlling a pest, a biological control agent requires a colonizing ability which allows it to keep pace with the spatial and temporal disruption of the habitat. Control is greatest if the agent has temporal persistence, so that it can maintain its population even in the temporary absence of the target species, and if it is an opportunistic forager, enabling it to rapidly exploit a pest population.

Joseph Needham noted a Chinese text dating from 304 AD, *Records of the Plants and Trees of the Southern Regions*, by Hsi Han, which describes mandarin oranges protected by large reddish-yellow citrus ants which attack and kill insect pests of the orange trees. The citrus ant (*Oecophylla smaragdina*) was rediscovered in the 20th century, and since 1958 has been used in China to protect orange groves.

One of the earliest successes in the west was in controlling *Icerya purchasi* (cottony cushion scale) in Australia, using a predatory insect *Rodolia cardinalis* (the vedalia beetle). This success was repeated in California using the beetle and a parasitoid fly, *Cryptochaetum iceryae*.

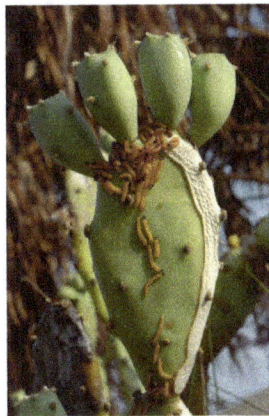

Cactoblastis cactorum larvae feeding on Opuntia cacti

Prickly pear cacti were introduced into Queensland, Australia as ornamental plants. They quickly spread to cover over 25 million hectares of Australia. Two control agents were used to help control the spread of the plant, the cactus moth *Cactoblastis cactorum*, and *Dactylopius* scale insects.

Damage from *Hypera postica*, the alfalfa weevil, a serious introduced pest of forage, was substantially reduced by the introduction of natural enemies. 20 years after their introduction the population of weevils in the alfalfa area treated for alfalfa weevil in the Northeastern United States remained 75 percent down.

The invasive species Alternanthera philoxeroides (alligator weed) was controlled in Florida (U.S.) by introducing alligator weed flea beetle.

Alligator weed was introduced to the United States from South America. It takes root in shallow water, interfering with navigation, irrigation, and flood control. The alligator weed flea beetle and two other biological controls were released in Florida, enabling the state to ban the use of herbicides to control alligator weed three years later. Another aquatic weed, the giant salvinia (*Salvinia molesta*) is a serious pest, covering waterways, reducing water flow and harming native species. Control with the salvinia weevil (*Cyrtobagous salviniae*) is effective in warm climates, and in Zimbabwe, a 99% control of the weed was obtained over a two-year period.

Small commercially reared parasitoidal wasps, *Trichogramma ostriniae*, provide limited and erratic control of the European corn borer (*Ostrinia nubilalis*), a serious pest. Careful formulations of the bacterium *Bacillus thuringiensis* are more effective.

The population of *Levuana iridescens*, the Levuana moth, a serious coconut pest in Fiji, was brought under control by a classical biological control program in the 1920s.

Augmentation

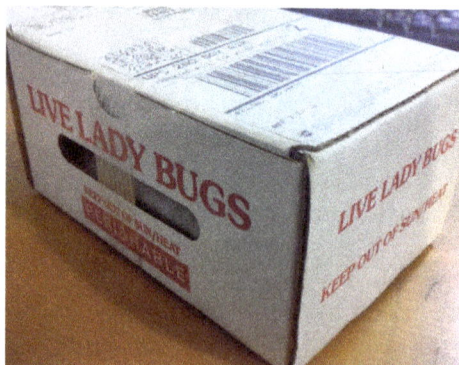

Hippodamia convergens, the convergent lady beetle, is commonly sold for biological control of aphids.

Augmentation involves the supplemental release of natural enemies, boosting the naturally occurring population. In inoculative release, small numbers of the control agents are released at intervals to allow them to reproduce, in the hope of setting up longer-term control, and thus keeping the pest down to a low level, constituting prevention rather than cure. In inundative release, in contrast, large numbers are released in the hope of rapidly reducing a damaging pest population, correcting a problem that has already arisen. Augmentation can be effective, but is not guaranteed to work, and relies on understanding of the situation.

An example of inoculative release occurs in greenhouse production of several crops. Periodic releases of the parasitoid, *Encarsia formosa*, are used to control greenhouse whitefly, while the predatory mite *Phytoseiulus persimilis* is used for control of the two-spotted spider mite.

The egg parasite *Trichogramma* is frequently released inundatively to control harmful moths. Similarly, *Bacillus thuringiensis* and other microbial insecticides are similarly used in large enough quantities for a rapid effect. Recommended release rates for *Trichogramma* in vegetable or field crops range from 5,000 to 200,000 per acre (1 to 50 per square metre) per week according to the level of pest infestation. Similarly, entomopathogenic nematodes are released at rates of millions and even billions per acre for control of certain soil-dwelling insect pests.

Conservation

The conservation of existing natural enemies in an environment is the third method of biological pest control. Natural enemies are already adapted to the habitat and to the target pest, and their conservation can be simple and cost-effective, as when nectar-producing crop plants are grown in the borders of rice fields. These provide nectar to support parasitoids and predators of planthopper pests and have been demonstrated to be so effective (reducing pest densities by 10- or even 100-fold) that farmers sprayed 70% less insecticides, enjoyed yields boosted by 5%, and this led to an economic advantage of 7.5%. Predators of aphids were similarly found to be present in tussock grasses by field boundary hedges in England, but they spread too slowly to reach the centres of fields. Control was improved by planting a metre-wide strip of tussock grasses in field centres, enabling aphid predators to overwinter there.

Cropping systems can be modified to favor natural enemies, a practice sometimes referred to as habitat manipulation. Providing a suitable habitat, such as a shelterbelt, hedgerow, or beetle bank where beneficial insects can live and reproduce, can help ensure the survival of populations of natural enemies. Things as simple as leaving a layer of fallen leaves or mulch in place provides a suitable food source for worms and provides a shelter for insects, in turn being a food source for such beneficial mammals as hedgehogs and shrews. Compost piles and stacks of wood can provide shelter for invertebrates and small mammals. Long grass and ponds support amphibians. Not removing dead annuals and non-hardy plants in the autumn allows insects to make use of their hollow stems during winter. In California, prune trees are sometimes planted in grape vineyards to provide an improved overwintering habitat or refuge for a key grape pest parasitoid. The providing of artificial shelters in the form of wooden caskets, boxes or flowerpots is also sometimes undertaken, particularly in gardens, to make a cropped area more attractive to natural enemies. For example, earwigs are natural predators which can be encouraged in gardens by hanging upside-down flowerpots filled with straw or wood wool. Green lacewings can be encouraged by using plastic bottles with an open bottom and a roll of cardboard inside. Birdhouses enable insectivorous birds to nest; the most useful birds can be attracted by choosing an opening just large enough for the desired species.

An inverted flowerpot filled with straw to attract earwigs

Biological Control Agents

Predators

Lacewings are available from biocontrol dealers.

Predators are mainly free-living species that directly consume a large number of prey during their whole lifetime. Ladybugs, and in particular their larvae which are active between May and July in the northern hemisphere, are voracious predators of aphids, and also consume mites, scale insects and small caterpillars. The spotted lady beetle (*Coleomegilla maculata*) is also able to feed on the eggs and larvae of the Colorado potato beetle (*Leptinotarsa decemlineata*).

The larvae of many hoverfly species principally feed upon greenfly (aphids), one larva devouring up to 400 in its lifetime. Their effectiveness in commercial crops has not been studied.

Predatory Polistes wasp looking for bollworms or other caterpillars on a cotton plant

Several species of entomopathogenic nematode are important predators of insect and other invertebrate pests. *Phasmarhabditis hermaphrodita* is a microscopic nematode that kills slugs. Its complex life cycle include a free-living, infective stage in the soil where it becomes associated with a pathogenic bacteria such as *Moraxella osloensis*. The nematode enters the slug through the posterior mantle region, thereafter feeding and reproducing inside, but it is the bacteria that kill the slug. The nematode is available commercially in Europe and is applied by watering onto moist soil.

Species used to control spider mites include the predatory mites *Phytoseiulus persimilis*, *Neoseilus californicus,* and *Amblyseius cucumeris*, the predatory midge *Feltiella acarisuga*, and a ladybird *Stethorus punctillum*. The bug *Orius insidiosus* has been successfully used against the two-spotted spider mite and the western flower thrips (*Frankliniella occidentalis*).

Parasitoids

Parasitoids lay their eggs on or in the body of an insect host, which is then used as a food for developing larvae. The host is ultimately killed. Most insect parasitoids are wasps or flies, and may have a very narrow host range. The most important groups are the ichneumonid wasps, which prey mainly on caterpillars of butterflies and moths; braconid wasps, which attack caterpillars and a wide range of other insects including greenfly; chalcid wasps, which parasitize eggs and larvae of greenfly, whitefly, cabbage caterpillars, and scale insects; and tachinid flies, which parasitize a wide range of insects including caterpillars, adult and larval beetles, and true bugs.

Encarsia formosa was one of the first biological control agents developed.

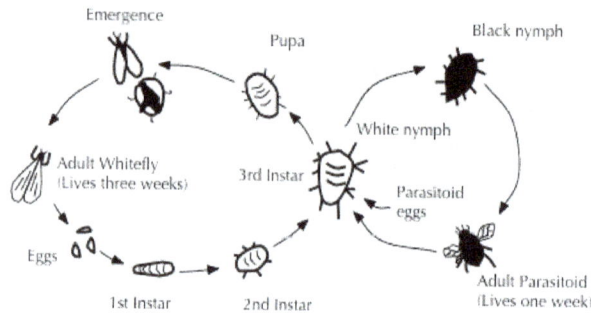

Life cycles of Greenhouse whitefly and its parasitoid wasp Encarsia formosa

Encarsia formosa is a small predatory chalcid wasp which is a parasitoid of white-fly, a sap-feeding insect which can cause wilting and black sooty moulds in glasshouse vegetable and ornamental crops. It is most effective when dealing with low level infestations, giving protection over a long period of time. The wasp lays its eggs in young whitefly 'scales', turning them black as the parasite larvae pupates. *Gonatocerus ashmeadi* (Hymenoptera: Mymaridae) has been introduced to control the glassy-winged sharpshooter *Homalodisca vitripennis* (Hemipterae: Cicadellidae) in French Polynesia and has successfully controlled ~95% of the pest density.

Parasitoids are among the most widely used biological control agents. Commercially, there are two types of rearing systems: short-term daily output with high production of parasitoids per day, and long-term low daily output with a range in production of 4-1000million female parasitoids per week. Larger production facilities produce on a yearlong basis, whereas some facilities produce only seasonally. Rearing facilities are usually a significant distance from where the agents are to be used in the field, and transporting the parasitoids from the point of production to the point of use can pose problems. Shipping conditions can be too hot, and even vibrations from planes or trucks can adversely affect parasitoids.

Pathogens

Pathogenic micro-organisms include bacteria, fungi, and viruses. They kill or debilitate their host and are relatively host-specific. Various microbial insect diseases occur naturally, but may also be used as biological pesticides. When naturally occurring, these outbreaks are density-dependent in that they generally only occur as insect populations become denser.

Bacteria

Bacteria used for biological control infect insects via their digestive tracts, so they offer only limited options for controlling insects with sucking mouth parts such as aphids and scale insects. *Bacillus thuringiensis* is the most widely applied species of bacteria used for biological control, with at least four sub-species used against Lepidopteran (moth, butterfly), Coleopteran (beetle) and Dipteran (true fly) insect pests. The bacterium is available in sachets of dried spores which are mixed with water and sprayed onto vulnerable plants such as brassicas and fruit trees. *B. thuringiensis* has also been incorporated into crops, making them resistant to these pests and thus reducing the use of pesticides. The bacterium *Paenibacillus popilliae* causes milky spore disease has been found useful in the control of Japanese beetle, killing the larvae. It is very specific to its host species and is harmless to vertebrates and other invertebrates.

Fungi

Green peach aphid, a pest in its own right and a vector of plant viruses, killed by the fungus Pandora neoaphidis (Zygomycota: Entomophthorales) Scale bar = 0.3 mm.

Entomopathogenic fungi, which cause disease in insects, include at least 14 species that attack aphids. *Beauveria bassiana* is mass-produced and used to manage a wide variety of insect pests including whiteflies, thrips, aphids and weevils. *Lecanicillium* spp. are deployed against white flies, thrips and aphids. *Metarhizium* spp. are used against pests including beetles, locusts and other grasshoppers, Hemiptera, and spider mites. *Paecilomyces fumosoroseus* is effective against white flies, thrips and aphids; *Purpureocillium lilacinus* is used against root-knot nematodes, and 89 *Trichoderma* species against certain plant pathogens. *Trichoderma viride* has been used against Dutch elm disease, and has shown some effect in suppressing silver leaf, a disease of stone fruits caused by the pathogenic fungus *Chondrostereum purpureum*.

The fungi *Cordyceps* and *Metacordyceps* are deployed against a wide spectrum of arthropods. *Entomophaga* is effective against pests such as the green peach aphid.

Several members of Chytridiomycota and Blastocladiomycota have been explored as agents of biological control. From Chytridiomycota, *Synchytrium solstitiale* is being considered as a control agent of the yellow star thistle (*Centaurea solstitialis*) in the United States.

Viruses

Baculoviruses are specific to individual insect host species and have been shown to be useful in biological pest control. For example, the Lymantria dispar multicapsid nuclear polyhedrosis virus has been used to spray large areas of forest in North America where larvae of the gypsy moth are causing serious defoliation. The moth larvae are killed by the virus they have eaten and die, the disintegrating cadavers leaving virus particles on the foliage to infect other larvae.

A mammalian virus, the rabbit haemorrhagic disease virus has been introduced to Australia and to New Zealand to attempt to control the European rabbit populations there.

Algae

Lagenidium giganteum is a water-borne mould that parasitizes the larval stage of mosquitoes. When applied to water, the motile spores avoid unsuitable host species and search out suitable mosquito larval hosts. This alga has the advantages of a dormant phase, resistant to desiccation, with slow-release characteristics over several years. Unfortunately, it is susceptible to many chemicals used in mosquito abatement programmes.

Plants

The legume vine *Mucuna pruriens* is used in the countries of Benin and Vietnam as a biological control for problematic *Imperata cylindrica* grass. *Mucuna pruriens* is said not to be invasive outside its cultivated area. *Desmodium uncinatum* can be used in push-pull farming to stop the parasitic plant, *Striga*.

Other Methods

Combined Use of Parasitoids and Pathogens

In cases of massive and severe infection of invasive pests, techniques of pest control are often used in combination. An example is the emerald ash borer, *Agrilus planipennis*, an invasive beetle from China, which has destroyed tens of millions of ash trees in its introduced range in North America. As part of the campaign against it, from 2003 American scientists and the Chinese Academy of Forestry searched for its natural enemies in the wild, leading to the discovery of several parasitoid wasps, namely *Tetrastichus planipennisi*, a gregarious larval endoparasitoid,*Oobius agrili*, a solitary, parthenogenic egg parasitoid, and *Spathius agrili*, a gregarious larval ectoparasitoid. These have been introduced and released into the United States of America as a possible biological control of the emerald ash borer. Initial results have shown promise with *Tetrastichus planipennisi* and it is now being released along with *Beauveria bassiana*, a fungal pathogen with known insecticidal properties.

Indirect Control

Pests may be controlled by biological control agents that do not prey directly upon them. For example, the Australian bush fly, *Musca vetustissima*, is a major nuisance pest in Australia, but native decomposers found in Australia are not adapted to feeding on cow dung, which is where bush flies breed. Therefore, the Australian Dung Beetle Project (1965–1985), led by Dr. George Bornemissza of the Commonwealth Scientific and Industrial Research Organisation, released forty-nine species of dung beetle, with the aim of reducing the amount of dung and therefore also the potential breeding sites of the fly.

Side-effects

Biological control can affect biodiversity through predation, parasitism, pathogenicity, competition, or other attacks on non-target species. An introduced control does not always target only the intended pest species; it can also target native species. In Hawaii during the 1940s parasitic wasps were introduced to control a lepidopteran pest and the wasps are still found there today. This may have a negative impact on the native ecosystem, however, host range and impacts need to be studied before declaring their impact on the environment.

Vertebrate animals tend to be generalist feeders, and seldom make good biological control agents; many of the classic cases of "biocontrol gone awry" involve vertebrates. For example, the cane toad (*Bufo marinus*) was intentionally introduced to Australia to control the greyback cane beetle (*Dermolepida albohirtum*), and other pests of sugar cane. 102 toads were obtained from Hawaii and bred in captivity to increase their numbers until they were released into the sugar cane fields of the tropic north in 1935. It was later discovered that the toads could not jump very high and so were unable to eat the cane beetles which stayed up on the upper stalks of the cane plants. However the toad thrived by feeding on other insects and it soon spread very rapidly; it took over native amphibian habitat and brought foreign disease to native toads and frogs, dramatically reducing their populations. Also when it is threatened or handled, the cane toad releases poison from parotoid glands on its shoulders; native Australian species such as goannas, tiger snakes, dingos and northern quolls that attempted to eat the toad were harmed or killed. However, there has been some recent evidence that native predators are adapting, both physiologically and through changing their behaviour, so in the long run, their populations may recover.

Rhinocyllus conicus, a seed-feeding weevil, was introduced to North America to control exotic musk thistle (*Carduus nutans*) and Canadian thistle (*Cirsium arvense*). However the weevil also attacks native thistles, harming such species as the endemic Platte thistle (*Cirsium neomexicanum*) by selecting larger plants (which reduced the gene pool), reducing seed production and ultimately threatening the species' survival.

The small Asian mongoose (*Herpestus javanicus*) was introduced to Hawaii in order to control the rat population. However it was diurnal and the rats emerged at night, and it preyed on the endemic birds of Hawaii, especially their eggs, more often than it ate the rats, and now both rats and mongooses threaten the birds. This introduction was undertaken without understanding the consequences of such an action. No regulations existed at the time, and more careful evaluation should prevent such releases now.

The sturdy and prolific eastern mosquitofish (*Gambusia holbrooki*) is a native of the southeastern United States and was introduced around the world in the 1930s and 40s to feed on mosquito larvae and thus combat malaria. However, it has thrived at the expense of local species, causing a decline of endemic fish and frogs through competition for food resources, as well as through eating their eggs and larvae. In Australia, the mosquitofish is the subject of discussion as to how best to control it; in 1989 it was said that "biological population control is well beyond present capabilities", and this remains the position.

Grower Education

A potential obstacle to the adoption of biological pest control measures is growers sticking to the familiar use of pesticides. It has been claimed that many of the pests that are controlled today using pesticides, actually became pests because pesticide use reduced or eliminated natural predators. A method of increasing grower adoption of biocontrol involves is letting growers learn by doing, for example showing them simple field experiments, having observations of live predation of pests, or collections of parasitised pests. In the Philippines, early season sprays against leaf folder caterpillars were common practice, but growers were asked to follow a 'rule of thumb' of not spraying against leaf folders for the first 30 days after transplanting; participation in this resulted in a reduction of insecticide use by 1/3 and a change in grower perception of insecticide use.

Beekeeping

Apiculture (from Latin: *apis* "bee") is the maintenance of honey bee colonies, commonly in hives, by humans. A beekeeper (or apiarist) keeps bees in order to collect their honey and other products that the hive produces (including beeswax, propolis, pollen, and royal jelly), to pollinate crops, or to produce bees for sale to other beekeepers. A location where bees are kept is called an apiary or "bee yard".

Beekeeping, tacuinum sanitatis casanatensis (14th century)

Beekeeping in Serbia

Depictions of humans collecting honey from wild bees date to 15,000 years ago. Beekeeping in pottery vessels began about 9,000 years ago in North Africa. Domestication is shown in Egyptian art from around 4,500 years ago. Simple hives and smoke were used and honey was stored in jars, some of which were found in the tombs of pharaohs such as Tutankhamun. It wasn't until the 18th century that European understanding of the colonies and biology of bees allowed the construction of the moveable comb hive so that honey could be harvested without destroying the entire colony.

Honey seeker depicted on 8000-year-old cave painting near Valencia, Spain

History of Beekeeping

At some point humans began to attempt to domesticate wild bees in artificial hives made from hollow logs, wooden boxes, pottery vessels, and woven straw baskets or "skeps". Traces of beeswax are found in pot sherds throughout the Middle East beginning about 7000 BCE.

Honeybees were kept in Egypt from antiquity. On the walls of the sun temple of Nyuserre Ini from the Fifth Dynasty, before 2422 BCE, workers are depicted blowing smoke into hives as they are removing honeycombs. Inscriptions detailing the production of honey are found on the tomb of Pabasa from the Twenty-sixth Dynasty (c. 650 BCE), depicting pouring honey in jars and cylindrical hives. Sealed pots of honey were found in the grave goods of pharaohs such as Tutankhamun.

There was a documented attempt to introduce bees to dry areas of Mesopotamia in the 8th century BCE by Shamash-resh-uṣur, the governor of Mari and Suhu. His plans were detailed in a stele of 760 BCE:

Stele showing Shamash-resh-uṣur praying to the gods Adad and Ishtar with an inscription in Babylonian cuneiform.	I am Shamash-resh-uṣur , the governor of Suhu and the land of Mari. Bees that collect honey, which none of my ancestors had ever seen or brought into the land of Suhu, I brought down from the mountain of the men of Habha, and made them settle in the orchards of the town 'Gabbari-built it'. They collect honey and wax, and I know how to melt the honey and wax – and the gardeners know too. Whoever comes in the future, may he ask the old men of the town, (who will say) thus: "They are the buildings of Shamash-resh-uṣur, the governor of Suhu, who introduced honey bees into the land of Suhu." — *translated text from stele, (Dalley, 2002)*

In prehistoric Greece (Crete and Mycenae), there existed a system of high-status apiculture, as can be concluded from the finds of hives, smoking pots, honey extractors and other beekeeping paraphernalia in Knossos. Beekeeping was considered a highly valued industry controlled by beekeeping overseers—owners of gold rings depicting apiculture scenes rather than religious ones as they have been reinterpreted recently, contra Sir Arthur Evans.

Archaeological finds relating to beekeeping have been discovered at Rehov, a Bronze and Iron Age archaeological site in the Jordan Valley, Israel. Thirty intact hives, made of straw and unbaked clay, were discovered by archaeologist Amihai Mazar in the ruins of the city, dating from about 900 BCE. The hives were found in orderly rows, three high, in a manner that could have accommodated around 100 hives, held more than 1 million bees and had a potential annual yield of 500 kilograms of honey and 70 kilograms of beeswax, according to Mazar, and are evidence that an advanced honey industry existed in ancient Israel 3,000 years ago.

The Beekeepers, 1568, by Pieter Bruegel the Elder

In ancient Greece, aspects of the lives of bees and beekeeping are discussed at length by Aristotle. Beekeeping was also documented by the Roman writers Virgil, Gaius Julius Hyginus, Varro, and Columella.

Beekeeping has also been practiced in ancient China since antiquity. In the book "Golden Rules of Business Success" written by Fan Li (or Tao Zhu Gong) during the Spring and Autumn Period there are sections describing the art of beekeeping, stressing the importance of the quality of the wooden box used and how this can affect the quality of the honey.

The ancient Maya domesticated a separate species of stingless bee. The use of stingless bees is referred to as meliponiculture, named after bees of the tribe Meliponini—such as *Melipona quadrifasciata* in Brazil. This variation of bee keeping still occurs around the world today. For instance, in Australia, the stingless bee *Tetragonula carbonaria* is kept for production of their honey.

Origins

There are more than 20,000 species of wild bees. Many species are solitary (e.g., mason bees, leafcutter bees (Megachilidae), carpenter bees and other ground-nesting bees). Many others rear their young in burrows and small colonies (e.g., bumblebees and stingless bees). Some honey bees are wild e.g. the little honeybee (*Apis florea*), giant honeybee (*Apis dorsata*) and rock bee (*Apis laboriosa*). Beekeeping, or apiculture, is concerned with the practical management of the social species of honey bees, which live in large colonies of up to 100,000 individuals. In Europe and America the species universally managed by beekeepers is the Western honey bee (*Apis mellifera*). This species has several sub-species or regional varieties, such as the Italian bee (*Apis mellifera ligustica*), European dark bee (*Apis mellifera mellifera*), and the Carniolan honey bee (*Apis mellifera carnica*). In the tropics, other species of social bees are managed for honey production, including the Asiatic honey bee (*Apis cerana*).

All of the *Apis mellifera* sub-species are capable of inter-breeding and hybridizing. Many bee breeding companies strive to selectively breed and hybridize varieties to produce desirable qualities: disease and parasite resistance, good honey production, swarming behaviour reduction, prolific breeding, and mild disposition. Some of these hybrids are marketed under specific brand names, such as the Buckfast Bee or Midnite Bee. The advantages of the initial F1 hybrids produced by these crosses include: hybrid vigor, increased honey productivity, and greater disease resistance. The disadvantage is that in subsequent generations these advantages may fade away and hybrids tend to be very defensive and aggressive.

Wild Honey Harvesting

Wild bees' nest, suspended from a branch

Collecting honey from wild bee colonies is one of the most ancient human activities and is still practiced by aboriginal societies in parts of Africa, Asia, Australia, and South America. In Africa, honeyguide birds have evolved a mutualist relationship with humans, leading them to hives and participating in the feast. This suggests honey harvesting by humans may be of great antiquity. Some of the earliest evidence of gathering honey from wild colonies is from rock paintings, dating to around Upper Paleolithic (13,000 BCE). Gathering honey from wild bee colonies is usually done by subduing the bees with smoke and breaking open the tree or rocks where the colony is located, often resulting in the physical destruction of the nest.

Study of Honey Bees

It was not until the 18th century that European natural philosophers undertook the scientific study of bee colonies and began to understand the complex and hidden world of bee biology. Preeminent among these scientific pioneers were Swammerdam, René Antoine Ferchault de Réaumur, Charles Bonnet, and Francois Huber. Swammerdam and Réaumur were among the first to use a microscope and dissection to understand the internal biology of honey bees. Réaumur was among the first to construct a glass walled observation hive to better observe activities within hives. He observed queens laying eggs in open cells, but still had no idea of how a queen was fertilized; nobody had ever witnessed the mating of a queen and drone and many theories held that queens were "self-fertile," while others believed that a vapor or "miasma" emanating from the drones fertilized queens without direct physical contact. Huber was the first to prove by observation and experiment that queens are physically inseminated by drones outside the confines of hives, usually a great distance away.

Following Réaumur's design, Huber built improved glass-walled observation hives and sectional hives that could be opened like the leaves of a book. This allowed inspecting individual wax combs and greatly improved direct observation of hive activity. Although he went blind before he was twenty, Huber employed a secretary, Francois Burnens, to make daily observations, conduct careful experiments, and keep accurate notes over more than twenty years. Huber confirmed that a hive consists of one queen who is the mother of all the female workers and male drones in the colony. He was also the first to confirm that mating with drones takes place outside of hives and that queens are inseminated by a number of successive matings with male drones, high in the air at a great distance from their hive. Together, he and Burnens dissected bees under the microscope and were among the first to describe the ovaries and spermatheca, or sperm store, of queens as well as the penis of male drones. Huber is universally regarded as "the father of modern bee-science" and his "Nouvelles Observations sur Les Abeilles (or "New Observations on Bees") revealed all the basic scientific truths for the biology and ecology of honeybees.

Invention of the Movable Comb Hive

Rural beekeeping in the 16th century

Early forms of honey collecting entailed the destruction of the entire colony when the honey was harvested. The wild hive was crudely broken into, using smoke to suppress the bees, the honeycombs were torn out and smashed up — along with the eggs, larvae and honey they contained. The liquid honey from the destroyed brood nest was strained through a sieve or basket. This was destructive and unhygienic, but for hunter-gatherer societies this did not matter, since the honey was generally consumed immediately and there were always more wild colonies to exploit. But in settled societies the destruction of the bee colony meant the loss of a valuable resource; this drawback made beekeeping both inefficient and something of a "stop and start" activity. There could be no continuity of production and no possibility of selective breeding, since each bee colony was destroyed at harvest time, along with its precious queen.

During the medieval period abbeys and monasteries were centers of beekeeping, since beeswax was highly prized for candles and fermented honey was used to make alcoholic mead in areas of Europe where vines would not grow. The 18th and 19th centuries saw successive stages of a revolution in beekeeping, which allowed the bees themselves to be preserved when taking the harvest.

Intermediate stages in the transition from the old beekeeping to the new were recorded for example by Thomas Wildman in 1768/1770, who described advances over the
destructive old skep-based beekeeping so that the bees no longer had to be killed to
harvest the honey. Wildman for example fixed a parallel array of wooden bars across
the top of a straw hive or skep (with a separate straw top to be fixed on later) "so that
there are in all seven bars of deal" [in a 10-inch-diameter (250 mm) hive] "to which
the bees fix their combs." He also described using such hives in a multi-storey configuration, foreshadowing the modern use of supers: he described adding (at a proper time) successive straw hives below, and eventually removing the ones above when
free of brood and filled with honey, so that the bees could be separately preserved at
the harvest for a following season. Wildman also described a further development, using hives with "sliding frames" for the bees to build their comb, foreshadowing more
modern uses of movable-comb hives. Wildman's book acknowledged the advances in
knowledge of bees previously made by Swammerdam, Maraldi, and de Réaumur—he
included a lengthy translation of Réaumur's account of the natural history of bees—and
he also described the initiatives of others in designing hives for the preservation of
bee-life when taking the harvest, citing in particular reports from Brittany dating from
the 1750s, due to Comte de la Bourdonnaye. However, the forerunners of the modern
hives with movable frames that are mainly used today are considered the traditional
basket top bar (movable comb) hives of Greece, known as "Greek beehives". The oldest
testimony on their use dates back to 1669 although it is probable that their use is more
than 3000 years old.

Lorenzo Langstroth (1810–1895)

The 19th century saw this revolution in beekeeping practice completed through the per-
fection of the movable comb hive by the American Lorenzo Lorraine Langstroth. Langstroth was the first person to make practical use of Huber's earlier discovery that there
was a specific spatial measurement between the wax combs, later called *the bee space*,
which bees do not block with wax, but keep as a free passage. Having determined this
bee space (between 5 and 8 mm, or 1/4 to 3/8"), Langstroth then designed a series of
wooden frames within a rectangular hive box, carefully maintaining the correct space
between successive frames, and found that the bees would build parallel honeycombs

in the box without bonding them to each other or to the hive walls. This enables the beekeeper to slide any frame out of the hive for inspection, without harming the bees or the comb, protecting the eggs, larvae and pupae contained within the cells. It also meant that combs containing honey could be gently removed and the honey extracted without destroying the comb. The emptied honey combs could then be returned to the bees intact for refilling. Langstroth's book, *The Hive and Honey-bee*, published in 1853, described his rediscovery of the bee space and the development of his patent movable comb hive.

The invention and development of the movable-comb-hive fostered the growth of commercial honey production on a large scale in both Europe and the USA.

Evolution of Hive Designs

Bees at the hive entrance

Langstroth's design for movable comb hives was seized upon by apiarists and inventors on both sides of the Atlantic and a wide range of moveable comb hives were designed and perfected in England, France, Germany and the United States. Classic designs evolved in each country: Dadant hives and Langstroth hives are still dominant in the USA; in France the De-Layens trough-hive became popular and in the UK a British National hive became standard as late as the 1930s although in Scotland the smaller Smith hive is still popular. In some Scandinavian countries and in Russia the traditional trough hive persisted until late in the 20th century and is still kept in some areas. However, the Langstroth and Dadant designs remain ubiquitous in the USA and also in many parts of Europe, though Sweden, Denmark, Germany, France and Italy all have their own national hive designs. Regional variations of hive evolved to reflect the climate, floral productivity and the reproductive characteristics of the various subspecies of native honey bee in each bio-region.

The differences in hive dimensions are insignificant in comparison to the common factors in all these hives: they are all square or rectangular; they all use movable wooden frames; they all consist of a floor, brood-box, honey super, crown-board and roof. Hives have traditionally been constructed of cedar, pine, or cypress wood, but in recent years hives made from injection molded dense polystyrene have become increasingly important.

Honey-laden honeycomb in a wooden frame

Hives also use queen excluders between the brood-box and honey supers to keep the queen from laying eggs in cells next to those containing honey intended for consumption. Also, with the advent in the 20th century of mite pests, hive floors are often replaced for part of (or the whole) year with a wire mesh and removable tray.

Pioneers of Practical and Commercial Beekeeping

The 19th century produced an explosion of innovators and inventors who perfected the design and production of beehives, systems of management and husbandry, stock improvement by selective breeding, honey extraction and marketing. Preeminent among these innovators were:

Petro Prokopovych, used frames with channels in the side of the woodwork, these were packed side by side in boxes that were stacked one on top of the other. The bees travelling from frame to frame and box to box via the channels. The channels were similar to the cut outs in the sides of modern wooden sections (1814).

Jan Dzierżon, was the father of modern apiology and apiculture. All modern beehives are descendants of his design.

L. L. Langstroth, revered as the "father of American apiculture", no other individual has influenced modern beekeeping practice more than Lorenzo Lorraine Langstroth. His classic book *The Hive and Honey-bee* was published in 1853.

Moses Quinby, often termed 'the father of commercial beekeeping in the United States', author of *Mysteries of Bee-Keeping Explained*.

Amos Root, author of the A B C of Bee Culture, which has been continuously revised and remains in print. Root pioneered the manufacture of hives and the distribution of bee-packages in the United States.

A. J. Cook, author of The Bee-Keepers' Guide; or Manual of the Apiary, 1876.

Dr. C.C. Miller was one of the first entrepreneurs to actually make a living from apiculture. By 1878 he made beekeeping his sole business activity. His book, *Fifty Years Among the Bees*, remains a classic and his influence on bee management persists to this day.

Honey spinner

Major Francesco De Hruschka was an Italian military officer who made one crucial invention that catalyzed the commercial honey industry. In 1865 he invented a simple machine for extracting honey from the comb by means of centrifugal force. His original idea was simply to support combs in a metal framework and then spin them around within a container to collect honey as it was thrown out by centrifugal force. This meant that honeycombs could be returned to a hive undamaged but empty, saving the bees a vast amount of work, time, and materials. This single invention greatly improved the efficiency of honey harvesting and catalysed the modern honey industry.

Walter T. Kelley was an American pioneer of modern beekeeping in the early and mid-20th century. He greatly improved upon beekeeping equipment and clothing and went on to manufacture these items as well as other equipment. His company sold via catalog worldwide and his book, How to Keep Bees & Sell Honey, an introductory book of apiculture and marketing, allowed for a boom in beekeeping following World War II.

In the U.K. practical beekeeping was led in the early 20th century by a few men, pre-eminently Brother Adam and his Buckfast bee and R.O.B. Manley, author of many titles, including Honey Production in the British Isles and inventor of the Manley frame, still universally popular in the U.K. Other notable British pioneers include William Herrod-Hempsall and Gale.

Dr. Ahmed Zaky Abushady (1892–1955), was an Egyptian poet, medical doctor, bacteriologist and bee scientist who was active in England and in Egypt in the early part of the twentieth century. In 1919, Abushady patented a removable, standardized aluminum honeycomb. In 1919 he also founded The Apis Club in Benson, Oxfordshire, and its periodical Bee World, which was to be edited by Annie D. Betts and later by Dr. Eva Crane. The Apis Club was transitioned to the International Bee Research Association (IBRA). Its archives are held in the National Library of Wales. In Egypt in the 1930s, Abushady established The Bee Kingdom League and its organ, The Bee Kingdom.

In India, R. N. Mattoo was the pioneer worker in starting beekeeping with Indian honeybee, (*Apis cerana indica*) in early 1930s. Beekeeping with European honeybee, (*Apis mellifera*) was started by Dr. A. S. Atwal and his team members, O. P. Sharma and **N. P. Goyal** in Punjab in early 1960s. It remained confined to Punjab and Himachal Pradesh up to late 1970s. Later on in 1982, Dr. R. C. Sihag, working at Haryana Agricultural University, Hisar (Haryana), introduced and established this honeybee in Haryana and standardized its management practices for semi-arid-subtropical climates.On the basis of these practices, beekeeping with this honeybee could be extended to the rest of the country. Now beekeeping with *Apis mellifera* predominates in India.

Traditional Beekeeping

Wooden hives in Stripeikiai honeymaking museum, Lithuania

Beekeeping at Kawah Ijen Mountain, Indonesia

Fixed Comb Hives

A fixed comb hive is a hive in which the combs cannot be removed or manipulated for management or harvesting without permanently damaging the comb. Almost any hollow structure can be used for this purpose, such as a log gum, skep, wooden box, or a clay pot or tube. Fixed comb hives are no longer in common use in industrialized countries, and are illegal in places that require movable combs to inspect for problems such as varroa and American foulbrood. In many developing countries fixed comb hives are widely used and, because they can be made from any locally available material, are very inexpensive. Beekeeping using fixed comb hives is an essential part of the livelihoods of many communities in poor countries. The charity Bees for Development recognizes that local skills to manage bees in fixed comb hives are widespread in Africa, Asia, and South America. Internal size of fixed comb hives range from 32.7 liters (2000 cubic inches) typical of the clay tube hives used in Egypt to 282 liters (17209 cubic inches) for the Perone hive. Straw skeps, bee gums, and unframed box hives are unlawful in most

US states, as the comb and brood cannot be inspected for diseases. However, skeps are still used for collecting swarms by hobbyists in the UK, before moving them into standard hives. Quinby used box hives to produce so much honey that he saturated the New York market in the 1860s. His writings contain excellent advice for management of bees in fixed comb hives.

Modern Beekeeping

Top-bar Hives

Top bar hives have been widely adopted in Africa where they are used to keep tropical honeybee ecotypes. Their advantages include being light weight, adaptable, easy to harvest honey, and less stressful for the bees. Disadvantages include combs that are fragile and cannot usually be extracted and returned to the bees to be refilled and that they cannot easily be expanded for additional honey storage.

A growing number of amateur beekeepers are adopting various top-bar hives similar to the type commonly found in Africa. Top bar hives were originally used as a traditional beekeeping method in Greece and Vietnam with a history dating back over 2000 years. These hives have no frames and the honey-filled comb is not returned after extraction. Because of this, the production of honey is likely to be somewhat less than that of a frame and super based hive such as Langstroth or Dadant. Top bar hives are mostly kept by people who are more interested in having bees in their garden than in honey production per se. Some of the most well known top-bar hive designs are the Kenyan Top Bar Hive with sloping sides, the Tanzanian Top Bar Hive with straight sides, and Vertical Top Bar Hives, such as the Warre or "People's Hive" designed by Abbe Warre in the mid-1900s.

The initial costs and equipment requirements are typically much less than other hive designs. Scrap wood or #2 or #3 pine can often be used to build a nice hive. Top-bar hives also offer some advantages to interacting with the bees and the amount of weight that must be lifted is greatly reduced. Top-bar hives are being widely used in developing countries in Africa and Asia as a result of the Bees for Development program. Since 2011, a growing number of beekeepers in the U.S. are using various top-bar hives.

Horizontal Frame Hives

The De-Layens hive, Jackson Horizontal Hive, and various chest type hives are widely used in Spain, France, Ukraine, Belarus, Africa, and parts of Russia. They are a step up from fixed comb and top bar hives because they have movable frames that can be extracted. Their limitation is primarily that volume is fixed and not easily expanded. Honey has to be removed one frame at a time, extracted or crushed, and the empty frames returned to be refilled. Various horizontal hives have been adapted and widely used for commercial migratory beekeeping. The Jackson Horizontal Hive is particularly well adapted for tropical agriculture. The De-Layens hive is popular in parts of Spain.

Vertical Stackable Frame Hives

In the United States, the Langstroth hive is commonly used. The Langstroth was the first successful top-opened hive with movable frames. Many other hive designs are based on the principle of bee space first described by Langstroth. The Langstroth hive is a descendant of Jan Dzierzon's Polish hive designs. In the United Kingdom, the most common type of hive is the British National, which can hold Hoffman, British Standard or Manley frames. It is not unusual to see some other sorts of hive (Smith, Commercial, WBC, Langstroth, and Rose). Dadant and Modified Dadant hives are widely used in France and Italy where their large size is an advantage. Square Dadant hives - often called 12 frame Dadant or Brother Adam hives - are used in large parts of Germany and other parts of Europe by commercial beekeepers. The Rose hive is a modern design that attempts to address many of the flaws and limitations of other movable frame hives. The only significant weakness of the Rose design is that it requires 2 or 3 boxes as a brood nest which infers a large number of frames to be worked when managing the bees. The major advantage shared by these designs is that additional brood and honey storage space can be added via boxes of frames added to the hive. This also simplifies honey collection since an entire box of honey can be removed instead of removing one frame at a time.

Protective Clothing

Beekeepers often wear protective clothing to protect themselves from stings

Most beekeepers also wear some protective clothing. Novice beekeepers usually wear gloves and a hooded suit or hat and veil. Experienced beekeepers sometimes elect not to use gloves because they inhibit delicate manipulations. The face and neck are the most important areas to protect, so most beekeepers wear at least a veil. Defensive bees are attracted to the breath, and a sting on the face can lead to much more pain and swelling than a sting elsewhere, while a sting on a bare hand can usually be quickly removed by fingernail scrape to reduce the amount of venom injected.

The protective clothing is generally light colored (but not colorful) and of a smooth material. This provides the maximum differentiation from the colony's natural predators (such as bears and skunks) which tend to be dark-colored and furry.

'Stings' retained in clothing fabric continue to pump out an alarm pheromone that attracts aggressive action and further stinging attacks. Washing suits regularly, and rinsing gloved hands in vinegar minimizes attraction.

Smoker

Bee smoker with heat shield and hook

Smoke is the beekeeper's third line of defense. Most beekeepers use a "smoker"—a device designed to generate smoke from the incomplete combustion of various fuels. Smoke calms bees; it initiates a feeding response in anticipation of possible hive abandonment due to fire. Smoke also masks alarm pheromones released by guard bees or when bees are squashed in an inspection. The ensuing confusion creates an opportunity for the beekeeper to open the hive and work without triggering a defensive reaction. In addition, when a bee consumes honey the bee's abdomen distends, supposedly making it difficult to make the necessary flexes to sting, though this has not been tested scientifically.

Smoke is of questionable use with a swarm, because swarms do not have honey stores to feed on in response. Usually smoke is not needed, since swarms tend to be less defensive, as they have no stores or brood to defend, and a fresh swarm has fed well from the hive.

Many types of fuel can be used in a smoker as long as it is natural and not contaminated with harmful substances. These fuels include hessian, twine, burlap, pine needles, corrugated cardboard, and mostly rotten or punky wood. Indian beekeepers, especially in Kerala, often use coconut fibers as they are readily available, safe, and of negligible expense. Some beekeeping supply sources also sell commercial fuels like pulped paper and compressed cotton, or even aerosol cans of smoke. Other beekeepers use sumac as fuel because it ejects lots of smoke and doesn't have an odor.

Some beekeepers are using "liquid smoke" as a safer, more convenient alternative. It is a water-based solution that is sprayed onto the bees from a plastic spray bottle.

Torpor may also be induced by the introduction of chilled air into the hive – while chilled carbon dioxide may have harmful long-term effects.

Effects of Stings and of Protective Measures

Some beekeepers believe that the more stings a beekeeper receives, the less irritation each causes, and they consider it important for safety of the beekeeper to be stung a few times a season. Beekeepers have high levels of antibodies (mainly IgG) reacting to the major antigen of bee venom, phospholipase A2 (PLA). Antibodies correlate with the frequency of bee stings.

The entry of venom into the body from bee-stings may also be hindered and reduced by protective clothing that allows the wearer to remove stings and venom sacs with a simple tug on the clothing. Although the stinger is barbed, a worker bee is less likely to become lodged into clothing than human skin.

If a beekeeper is stung by a bee, there are many protective measures that should be taken in order to make sure the affected area does not become too irritated. The first cautionary step that should be taken following a bee sting is removing the stinger without squeezing the attached venom glands. A quick scrape with a fingernail is effective and intuitive. This step is effective in making sure that the venom injected does not spread, so the side effects of the sting will go away sooner. Washing the affected area with soap and water is also a good way to stop the spread of venom. The last step that needs to be taken is to apply ice or a cold compress to the stung area.

Natural Beekeeping

The natural beekeeping movement believes that modern beekeeping and agricultural practices, such as crop spraying, hive movement, frequent hive inspections, artificial insemination of queens, routine medication, and sugar water feeding, weaken bee hives.

Practitioners of 'natural beekeeping' tend to use variations of the top-bar hive, which is a simple design that retains the concept of movable comb without the use of frames or foundation. The horizontal top-bar hive, as championed by Marty Hardison, Michael Bush, Philip Chandler, Dennis Murrell and others, can be seen as a modernization of hollow log hives, with the addition of wooden bars of specific width from which bees hang their combs. Its widespread adoption in recent years can be attributed to the publication in 2007 of *The Barefoot Beekeeper* by Philip Chandler, which challenged many aspects of modern beekeeping and offered the horizontal top-bar hive as a viable alternative to the ubiquitous Langstroth-style movable-frame hive.

The most popular vertical top-bar hive is probably the Warré hive, based on a design by the French priest Abbé Émile Warré (1867–1951) and popularized by Dr. David Heaf in his English translation of Warré's book *L'Apiculture pour Tous* as *Beekeeping For All.*

Urban or Backyard Beekeeping

Related to natural beekeeping, urban beekeeping is an attempt to revert to a less industrialized way of obtaining honey by utilizing small-scale colonies that pollinate urban gardens. Urban apiculture has undergone a renaissance in the first decade of the 21st century, and urban beekeeping is seen by many as a growing trend.

Honey bee in Toronto

Some have found that "city bees" are actually healthier than "rural bees" because there are fewer pesticides and greater biodiversity. Urban bees may fail to find forage, however, and homeowners can use their landscapes to help feed local bee populations by planting flowers that provide nectar and pollen. An environment of year-round, uninterrupted bloom creates an ideal environment for colony reproduction.

Bee Colonies

Castes

A colony of bees consists of three castes of bee:

- a queen bee, which is normally the only breeding female in the colony;

- a large number of female worker bees, typically 30,000–50,000 in number;

- a number of male drones, ranging from thousands in a strong hive in spring to very few during dearth or cold season.

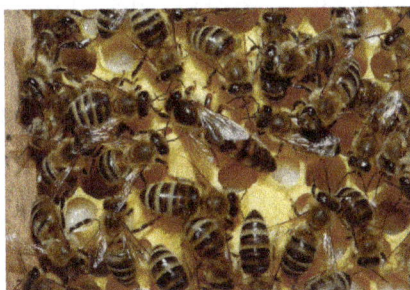
Queen bee (center)

The queen is the only sexually mature female in the hive and all of the female worker bees and male drones are her offspring. The queen may live for up to three years or more and may be capable of laying half a million eggs or more in her lifetime. At the peak of the breeding season, late spring to summer, a good queen may be capable of laying 3,000 eggs in one day, more than her own body weight. This would be exceptional however; a prolific queen might peak at 2,000 eggs a day, but a more average queen might lay just 1,500 eggs per day. The queen is raised from a normal worker egg, but is fed a larger amount of royal jelly than a normal worker bee, resulting in a radically different growth and metamorphosis. The queen influences the colony by the production and dissemination of a variety of pheromones or "queen substances". One of these chemicals suppresses the development of ovaries in all the female worker bees in the hive and prevents them from laying eggs.

Mating of Queens

The queen emerges from her cell after 15 days of development and she remains in the hive for 3–7 days before venturing out on a mating flight. Mating flight is otherwise known as 'nuptial flight'. Her first orientation flight may only last a few seconds, just enough to mark the position of the hive. Subsequent mating flights may last from 5 minutes to 30 minutes, and she may mate with a number of male drones on each flight. Over several matings, possibly a dozen or more, the queen receives and stores enough sperm from a succession of drones to fertilize hundreds of thousands of eggs. If she does not manage to leave the hive to mate—possibly due to bad weather or being trapped in part of the hive—she remains infertile and become a *drone layer*, incapable of producing female worker bees. Worker bees sometimes kill a non-performing queen and produce another. Without a properly performing queen, the hive is doomed.

Mating takes place at some distance from the hive and often several hundred feet in the air; it is thought that this separates the strongest drones from the weaker ones, ensuring that only the fastest and strongest drones get to pass on their genes.

Worker Bees

Almost all the bees in a hive are female worker bees. At the height of summer when activity in the hive is frantic and work goes on non-stop, the life of a worker bee may be as short as 6 weeks; in late autumn, when no brood is being raised and no nectar is being harvested, a young bee may live for 16 weeks, right through the winter.

Female worker bee

Over the course of their lives, worker bees' duties are dictated by age. For the first few weeks of their lifespan, they perform basic chores within the hive: cleaning empty brood cells, removing debris and other housekeeping tasks, making wax for building or repairing comb, and feeding larvae. Later, they may ventilate the hive or guard the entrance. Older workers leave the hive daily, weather permitting, to forage for nectar, pollen, water, and propolis.

Period	Work activity
Days 1-3	Cleaning cells and incubation
Day 3-6	Feeding older larvae
Day 6-10	Feeding younger larvae
Day 8-16	Receiving nectar and pollen from field bees
Day 12-18	Beeswax making and cell building
Day 14 onwards	Entrance guards; nectar, pollen, water and propolis foraging; robbing other hives

Drones

Larger drones compared to smaller workers

Drones are the largest bees in the hive (except for the queen), at almost twice the size of a worker bee. They do not work, do not forage for pollen or nectar, are unable to sting, and have no other known function than to mate with new queens and fertilize them on their mating flights. A bee colony generally starts to raise drones a few weeks before building queen cells so they can supersede a failing queen or prepare for swarming. When queen-raising for the season is over, bees in colder climates drive drones out of the hive to die, biting and tearing their legs and wings.

Differing Stages of Development

Stage of development	Queen	Worker	Drone
Egg	3 days	3 days	3 days
Larva	8 days	10 days	13 days :Successive moults occur within this period 8 to 13 day period

Cell Capped	day 8	day 8	day 10
Pupa	4 days	8 days	8 days
Total	15 days	21 days	24 days

Structure of a Bee Colony

A domesticated bee colony is normally housed in a rectangular hive body, within which eight to ten parallel frames house the vertical plates of honeycomb that contain the eggs, larvae, pupae and food for the colony. If one were to cut a vertical cross-section through the hive from side to side, the brood nest would appear as a roughly ovoid ball spanning 5-8 frames of comb. The two outside combs at each side of the hive tend to be exclusively used for long-term storage of honey and pollen.

Within the central brood nest, a single frame of comb typically has a central disk of eggs, larvae and sealed brood cells that may extend almost to the edges of the frame. Immediately above the brood patch an arch of pollen-filled cells extends from side to side, and above that again a broader arch of honey-filled cells extends to the frame tops. The pollen is protein-rich food for developing larvae, while honey is also food but largely energy rich rather than protein rich. The nurse bees that care for the developing brood secrete a special food called 'royal jelly' after feeding themselves on honey and pollen. The amount of royal jelly fed to a larva determines whether it develops into a worker bee or a queen.

Apart from the honey stored within the central brood frames, the bees store surplus honey in combs above the brood nest. In modern hives the beekeeper places separate boxes, called 'supers', above the brood box, in which a series of shallower combs is provided for storage of honey. This enables the beekeeper to remove some of the supers in the late summer, and to extract the surplus honey harvest, without damaging the colony of bees and its brood nest below. If all the honey is 'stolen', including the amount of honey needed to survive winter, the beekeeper must replace these stores by feeding the bees sugar or corn syrup in autumn.

Annual Cycle of a Bee Colony

The development of a bee colony follows an annual cycle of growth that begins in spring with a rapid expansion of the brood nest, as soon as pollen is available for feeding larvae. Some production of brood may begin as early as January, even in a cold winter, but breeding accelerates towards a peak in May (in the northern hemisphere), producing an abundance of harvesting bees synchronized to the main nectar flow in that region. Each race of bees times this build-up slightly differently, depending on how the flora of its original region blooms. Some regions of Europe have two nectar flows: one in late spring and another in late August. Other regions have only a single nectar flow. The skill of the beekeeper lies in predicting when the nectar flow will occur in his area and in trying to ensure that his colonies achieve a maximum population of harvesters at exactly the right time.

The key factor in this is the prevention or skillful management of the swarming impulse. If a colony swarms unexpectedly and the beekeeper does not manage to capture the resulting swarm, he is likely to harvest significantly less honey from that hive, since he has lost half his worker bees at a single stroke. If, however, he can use the swarming impulse to breed a new queen but keep all the bees in the colony together, he maximizes his chances of a good harvest. It takes many years of learning and experience to be able to manage all these aspects successfully, though owing to variable circumstances many beginners often achieve a good honey harvest.

Formation of New Colonies

Colony Reproduction: Swarming and Supersedure

A swarm about to land

All colonies are totally dependent on their queen, who is the only egg-layer. However, even the best queens live only a few years and one or two years longevity is the norm. She can choose whether or not to fertilize an egg as she lays it; if she does so, it develops into a female worker bee; if she lays an unfertilized egg it becomes a male drone. She decides which type of egg to lay depending on the size of the open brood cell she encounters on the comb. In a small worker cell, she lays a fertilized egg; if she finds a larger drone cell, she lays an unfertilized drone egg.

All the time that the queen is fertile and laying eggs she produces a variety of pheromones, which control the behavior of the bees in the hive. These are commonly called *queen substance*, but there are various pheromones with different functions. As the queen ages, she begins to run out of stored sperm, and her pheromones begin to fail. Inevitably, the queen begins to falter, and the bees decide to replace her by creating a new queen from one of her worker eggs. They may do this because she has been damaged (lost a leg or an antenna), because she has run out of sperm and cannot lay fertilized eggs (has become a 'drone laying queen'), or because her pheromones have dwindled to where they cannot control all the bees in the hive.

At this juncture, the bees produce one or more queen cells by modifying existing worker cells that contain a normal female egg. However, the bees pursue two distinct behaviors:

1. Supersedure: queen replacement within one hive without swarming

2.   Swarm cell production: the division of the hive into two colonies by swarming

Different sub-species of *Apis mellifera* exhibit differing swarming characteristics that reflect their evolution in different ecotopes of the European continent. In general the more northerly black races are said to swarm less and supersede more, whereas the more southerly yellow and grey varieties are said to swarm more frequently. The truth is complicated because of the prevalence of cross-breeding and hybridization of the sub species and opinions differ.

Supersedure is highly valued as a behavioral trait by beekeepers because a hive that supersedes its old queen does not swarm and so no stock is lost; it merely creates a new queen and allows the old one to fade away, or alternatively she is killed when the new queen emerges. When superseding a queen, the bees produce just one or two queen cells, characteristically in the center of the face of a broodcomb.

In swarming, by contrast, a great many queen cells are created—typically a dozen or more—and these are located around the edges of a broodcomb, most often at the sides and the bottom.

New wax combs between basement joists

Once either process has begun, the old queen normally leaves the hive with the hatching of the first queen cells. She leaves accompanied by a large number of bees, predominantly young bees (wax-secretors), who form the basis of the new hive. Scouts are sent out from the swarm to find suitable hollow trees or rock crevices. As soon as one is found, the entire swarm moves in. Within a matter of hours, they build new wax brood combs, using honey stores that the young bees have filled themselves with before leaving the old hive. Only young bees can secrete wax from special abdominal segments, and this is why swarms tend to contain more young bees. Often a number of virgin queens accompany the first swarm (the 'prime swarm'), and the old queen is replaced as soon as a daughter queen mates and begins laying. Otherwise, she is quickly superseded in the new home.

Factors That Trigger Swarming

It is generally accepted that a colony of bees does not swarm until they have completed all of their brood combs, i.e., filled all available space with eggs, larvae, and brood. This generally occurs in late spring at a time when the other areas of the hive are rapidly filling with honey stores. One key trigger of the swarming instinct is when the queen has no more room to lay eggs and the hive population is becoming very congested. Under these conditions, a prime swarm may issue with the queen, resulting in a halving of the population within the hive, leaving the old colony with a large number of hatching bees. The queen who leaves finds herself in a new hive with no eggs and no larvae but lots of energetic young bees who create a new set of brood combs from scratch in a very short time.

Another important factor in swarming is the age of the queen. Those under a year in age are unlikely to swarm unless they are extremely crowded, while older queens have swarming predisposition.

Beekeepers monitor their colonies carefully in spring and watch for the appearance of queen cells, which are a dramatic signal that the colony is determined to swarm.

When a colony has decided to swarm, queen cells are produced in numbers varying to a dozen or more. When the first of these queen cells is sealed after eight days of larval feeding, a virgin queen pupates and is due to emerge seven days later. Before leaving, the worker bees fill their stomachs with honey in preparation for the creation of new honeycombs in a new home. This cargo of honey also makes swarming bees less inclined to sting. A newly issued swarm is noticeably gentle for up to 24 hours and is often capable of being handled by a beekeeper without gloves or veil.

This swarm looks for shelter. A beekeeper may capture it and introduce it into a new hive, helping meet this need. Otherwise, it returns to a feral state, in which case it finds shelter in a hollow tree, excavation, abandoned chimney, or even behind shutters.

Back at the original hive, the first virgin queen to emerge from her cell immediately seeks to kill all her rival queens still waiting to emerge. Usually, however, the bees deliberately prevent her from doing this, in which case, she too leads a second swarm from the hive. Successive swarms are called 'after-swarms' or 'casts' and can be very small, often with just a thousand or so bees—as opposed to a prime swarm, which may contain as many as ten to twenty-thousand bees.

A swarm attached to a branch

A small after-swarm has less chance of survival and may threaten the original hive's survival if the number of individuals left is unsustainable. When a hive swarms despite the beekeeper's preventative efforts, a good management practice is to give the reduced hive a couple frames of open brood with eggs. This helps replenish the hive more quickly and gives a second opportunity to raise a queen if there is a mating failure.

Each race or sub-species of honey bee has its own swarming characteristics. Italian bees are very prolific and inclined to swarm; Northern European black bees have a strong tendency to supersede their old queen without swarming. These differences are the result of differing evolutionary pressures in the regions where each sub-species evolved.

Artificial Swarming

When a colony accidentally loses its queen, it is said to be "queenless". The workers realize that the queen is absent after as little as an hour, as her pheromones fade in the hive. The colony cannot survive without a fertile queen laying eggs to renew the population, so the workers select cells containing eggs aged less than three days and enlarge these cells dramatically to form "emergency queen cells". These appear similar to large peanut-like structures about an inch long that hang from the center or side of the brood combs. The developing larva in a queen cell is fed differently from an ordinary worker-bee; in addition to the normal honey and pollen, she receives a great deal of royal jelly, a special food secreted by young 'nurse bees' from the hypopharyngeal gland. This special food dramatically alters the growth and development of the larva so that, after metamorphosis and pupation, it emerges from the cell as a queen bee. The queen is the only bee in a colony which has fully developed ovaries, and she secretes a pheromone which suppresses the normal development of ovaries in all her workers.

Beekeepers use the ability of the bees to produce new queens to increase their colonies in a procedure called *splitting a colony*. To do this, they remove several brood combs from a healthy hive, taking care to leave the old queen behind. These combs must contain eggs or larvae less than three days old and be covered by young *nurse bees*, which care for the brood and keep it warm. These brood combs and attendant nurse bees are then placed into a small 'nucleus hive' with other combs containing honey and pollen. As soon as the nurse bees find themselves in this new hive and realize they have no queen, they set about constructing emergency queen cells using the eggs or larvae they have in the combs with them.

Diseases

The common agents of disease that affect adult honey bees include fungi, bacteria, protozoa, viruses, parasites, and poisons. The gross symptoms displayed by affected adult bees are very similar, whatever the cause, making it difficult for the apiarist to ascertain the causes of problems without microscopic identification of microorganisms or chemical analysis of poisons. Since 2006 colony losses from Colony Collapse Disorder have been increasing across the world although the causes of the syndrome are, as yet, unknown. In the US, commercial beekeepers have been increasing the number of hives to deal with higher rates attrition.

World Apiculture

World honey production and consumption in 2005				
Country	Production (1000 metric tons)	Consumption (1000 metric tons)	Number of beekeepers	Number of bee hives
Europe and Russia				
Ukraine	71.46	52		
Russia	52.13	54		
Spain	37.00	40		
Germany (*2008)	21.23	89	90,000*	1,000,000*
Hungary	19.71	4		
Romania	19.20	10		
Greece	16.27	16		
France	15.45	30		
Bulgaria	11.22	2		
Serbia	3 to 5	6.3	30,000	430,000
Denmark (*1996)	2.5	5	*4,000	*150,000
North America				
United States (*2006, **2002)	70.306*	158.75*	12,029** (210,000 bee keepers)	2,400,000*
Canada	45 (2006); 28 (2007)	29	13,000	500,000
Latin America				
Argentina	93.42 (Average 84)	3		
Mexico	50.63	31		
Brazil	33.75	2		
Uruguay	11.87	1		
Oceania				
Australia	18.46	16	12,000	520,000
New Zealand	9.69	8	2602	313,399
Asia				
China	299.33 (average 245)	238		7,200,000
Turkey	82.34 (average 70)	66		4,500,000
Iran				3,500,000
India	52.23	45		9,800,000
South Korea	23.82	27		
Vietnam	13.59	0		

World honey production and consumption in 2005				
Country	Production (1000 metric tons)	Consumption (1000 metric tons)	Number of beekeepers	Number of bee hives
Europe and Russia				
Turkmenistan	10.46	10		
Africa				
Ethiopia	41.23	40		4,400,000
Tanzania	28.68	28		
Angola	23.77	23		
Kenya	22.00	21		
Egypt (*1997)	16*		200,000*	2,000,000*
Central African Republic	14.23	14		
Morocco (*1997)	4.5*		27,000*	400,000*
South Africa (*2008)	~2.5*	~1.5*	~1,790*	~92,000*
Source: Food and Agriculture Organization of the United Nations				

Sources:

- Denmark: beekeeping.com (1996)

- Arab countries: beekeeping.com (1997)

- USA: University of Arkansas National Agricultural Law Center, Agricultural Marketing Resource Center

- Serbia

Sericulture

silkworm and cocoon

Sericulture, or silk farming, is the rearing of silkworms for the production of silk. Although there are several commercial species of silkworms, *Bombyx mori* (the caterpillar of the domesticated silk moth) is the most widely used and intensively studied silkworm. Silk was first produced in China as early as the Neolithic period. Sericulture has become an important cottage industry in countries such as Brazil, China, France, India, Italy, Japan, Korea, and Russia. Today, China and India are the two main producers, with more than 60% of the world's annual production.

History

According to Confucian text, the discovery of silk production dates to about 2700 BC, although archaeological records point to silk cultivation as early as the Yangshao period (5000 – 3000 BC). By about the first half of the 1st century AD it had reached ancient Khotan, by a series of interactions along the Silk Road. By 140 AD the practice had been established in India. In the 6th century the smuggling of silkworm eggs into the Byzantine Empire led to its establishment in the Mediterranean, remaining a monopoly in the Byzantine Empire for centuries (Byzantine silk). In 1147, during the Second Crusade, Roger II of Sicily (1095–1154) attacked Corinth and Thebes, two important centres of Byzantine silk production, capturing the weavers and their equipment and establishing his own silkworks in Palermo and Calabria, eventually spreading the industry to Western Europe.

The cocoons are weighed.

The cocoons are soaked and the silk is wound on spools.

The silk is woven using a loom.

Production

Silkworm larvae are fed with mulberry leaves, and, after the fourth moult, climb a twig placed near them and spin their silken cocoons. This process is achieved by the worm through a dense fluid secreted from its structural glands, resulting in the fiber of the cocoon. The silk is a continuous filament comprising fibroin protein, secreted from two salivary glands in the head of each larva, and a gum called sericin, which cements the filaments. The sericin is removed by placing the cocoons in hot water, which frees the silk filaments and readies them for reeling. This is known as the degumming process. The immersion in hot water also kills the silkworm pupae.

Single filaments are combined to form thread, which is drawn under tension through several guides and wound onto reels. The threads may be plied to form yarn. After drying, the raw silk is packed according to quality.

Stages of Production

The stages of production are as follows:

1. The silk moth lays thousands of eggs.

2. The silk moth eggs hatch to form larvae or caterpillars, known as silkworms.

3. The larvae feed on mulberry leaves.

4. Having grown and moulted several times silkworm weaves a net to hold itself

5. It swings its head from side to side in a figure '8' distributing the saliva that will form silk.

6. The silk solidifies when it contacts the air.

7. The silkworm spins approximately one mile of filament and completely encloses itself in a cocoon in about two or three days. The amount of usable quality silk in each cocoon is small. As a result, about 2500 silkworms are required to produce a pound of raw silk

8. The intact cocoons are boiled, killing the silkworm pupae.

9. The silk is obtained by brushing the undamaged cocoon to find the outside end of the filament.

10. The silk filaments are then wound on a reel. One cocoon contains approximately 1,000 yards of silk filament. The silk at this stage is known as raw silk. One thread comprises up to 48 individual silk filaments.

Mahatma Gandhi was critical of silk production based on the Ahimsa philosophy "not to hurt any living thing". He also promoted 'Ahimsa silk', made without boiling the pupae to procure the silk and wild silk made from the cocoons of wild and semi-wild silk moths. In the early 21st century the organisation PETA has campaigned against silk.

third stage of silkworm

silkworms on to Modern Rotary mountage

Biomimetics

Velcro tape mimics biological examples of multiple hooked structures such as burs.

Biomimetics or biomimicry is the imitation of the models, systems, and elements of nature for the purpose of solving complex human problems.

Living organisms have evolved well-adapted structures and materials over geological time through natural selection. Biomimetics has given rise to new technologies inspired by biological solutions at macro and nanoscales. Humans have looked at nature for answers to problems throughout our existence. Nature has solved engineering problems such as self-healing abilities, environmental exposure tolerance and resistance, hydrophobicity, self-assembly, and harnessing solar energy.

History

One of the early examples of biomimicry was the study of birds to enable human flight. Although never successful in creating a "flying machine", Leonardo da Vinci (1452–1519) was a keen observer of the anatomy and flight of birds, and made numerous notes and sketches on his observations as well as sketches of "flying machines". The Wright Brothers, who succeeded in flying the first heavier-than-air aircraft in 1903, derived inspiration from observations of pigeons in flight.

Biomimetics was coined by the American biophysicist and polymath Otto Schmitt during the 1950s. It was during his doctoral research that he developed the Schmitt trigger by studying the nerves in squid, attempting to engineer a device that replicated the biological system of nerve propagation. He continued to focus on devices that mimic natural systems and by 1957 he had perceived a converse to the standard view of biophysics at that time, a view he would come to call biomimetics.

Biophysics is not so much a subject matter as it is a point of view. It is an approach to problems of biological science utilizing the theory and technology of the physical sciences. Conversely, biophysics is also a biologist's approach to problems of physical science and engineering, although this aspect has largely been neglected.

— Otto Herbert Schmitt, In Appreciation, A Lifetime of Connections: Otto Herbert Schmitt, 1913 - 1998

A similar term, *Bionics* was coined by Jack E. Steele in 1960 at Wright-Patterson Air Force Base in Dayton, Ohio where Otto Schmitt also worked. Steele defined bionics as "the science of systems which have some function copied from nature, or which represent characteristics of natural systems or their analogues". During a later meeting in 1963 Schmitt stated,

Let us consider what bionics has come to mean operationally and what it or some word like it (I prefer biomimetics) ought to mean in order to make good use of the technical skills of scientists specializing, or rather, I should say, despecializing into this area of research

— Otto Herbert Schmitt, In Appreciation, A Lifetime of Connections: Otto Herbert Schmitt, 1913 - 1998

In 1969 the term biomimetics was used by Schmitt to title one of his papers, and by 1974 it had found its way into Webster's Dictionary, bionics entered the same dictionary earlier in 1960 as "a science concerned with the application of data about the functioning of biological systems to the solution of engineering problems". Bionic took on a different connotation when Martin Caidin referenced Jack Steele and his work in the novel *Cyborg* which later resulted in the 1974 television series *The Six Million Dollar Man* and its spin-offs. The term bionic then became associated with "the use of electronically operated artificial body parts" and "having ordinary human powers increased by or as if by the aid of such devices". Because the term *bionic* took on the implication of supernatural strength, the scientific community in English speaking countries largely abandoned it.

Velcro was inspired by the tiny hooks found on the surface of burs.

The term *biomimicry* appeared as early as 1982. Biomimicry was popularized by scientist and author Janine Benyus in her 1997 book *Biomimicry: Innovation Inspired by Nature*. Biomimicry is defined in the book as a "new science that studies nature's models and then imitates or takes inspiration from these designs and processes to solve human problems". Benyus suggests looking to Nature as a "Model, Measure, and Mentor" and emphasizes sustainability as an objective of biomimicry.

Existing Commercialized Applications

Fabrication

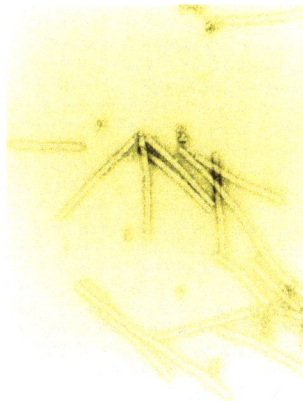

Scanning electron micrograph of rod shaped tobacco mosaic virus particles

Biomorphic mineralization is a technique that produces materials with morphologies and structures resembling those of natural living organisms by using bio-structures as templates for mineralization. Compared to other methods of material production, biomorphic mineralization is facile, environmentally benign and economic.

Display Technology

Vibrant blue color of *Morpho* butterfly due to structural coloration

Morpho butterfly wings contain microstructures that create its coloring effect through structural coloration rather than pigmentation. Incident light waves are reflected at specific wavelengths to create vibrant colors due to multilayer interference, diffraction, thin film interference, and scattering properties. The scales of these butterflies consist of microstructures such as ridges, cross-ribs, ridge-lamellae, and microribs that have been shown to be responsible for coloration. The structural color has been simply explained as the interference due to alternating layers of cuticle and air using a model of multilayer interference. The same principles behind the coloration of soap bubbles apply to butterfly wings. The color of butterfly wings is due to multiple instances of constructive interference from structures such as this. The photonic microstructure of butterfly wings can be replicated through biomorphic mineralization to yield similar properties. The photonic microstructures can be replicated using metal oxides or metal alkoxides such as titanium sulfate ($TiSO_4$), zirconium oxide (ZrO_2), and aluminium oxide (Al_2O_3). An alternative method of vapor-phase oxidation of SiH4 on the template surface was found to preserve delicate structural features of the microstructure. A display technology ("Mirasol") based on the reflective properties of *Morpho* butterfly wings was commercialized by Qualcomm in 2007. The technology uses Interferometric Modulation to reflect light so only the desired color is visible in each individual pixel of the display.

Possible Future Applications

Biomimetics

Biomimetics could in principle be applied in many fields. Because of the complexity of biological systems, the number of features that might be imitated is large. Biomimetic applications are at various stages of development from technologies that might become commercially usable to prototypes.

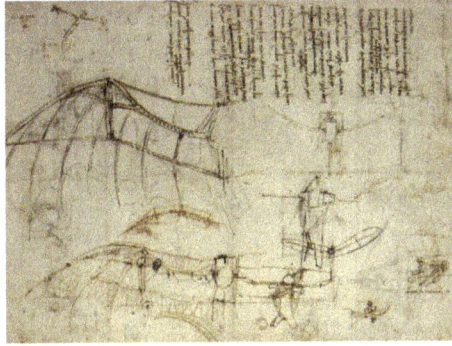

Leonardo da Vinci's design for a flying machine with wings based closely upon the structure of bat wings

Prototypes

Researchers studied the termite's ability to maintain virtually constant temperature and humidity in their termite mounds in Africa despite outside temperatures that vary from 1.5 °C to 40 °C (35 °F to 104 °F). Researchers initially scanned a termite mound and created 3-D images of the mound structure, which revealed construction that could influence human building design. The Eastgate Centre, a mid-rise office complex in Harare, Zimbabwe, stays cool without air conditioning and uses only 10% of the energy of a conventional building of the same size.

In structural engineering, the Swiss Federal Institute of Technology (EPFL) has incorporated biomimetic characteristics in an adaptive deployable "tensegrity" bridge. The bridge can carry out self-diagnosis and self-repair.

Technologies

Practical underwater adhesion is an engineering challenge since current technology is unable to stick surface strongly underwater because of barriers such as hydration layers and contaminants on surfaces. However, marine mussels can stick easily and efficiently to surfaces underwater under the harsh conditions of the ocean. They use strong filaments to adhere to rocks in the inter-tidal zones of wave-swept beaches, preventing them from being swept away in strong sea currents. Mussel foot proteins attach the filaments to rocks, boats and practically any surface in nature including other mussels. These proteins contain a mix of amino acid residues which has been adapted specifically for adhesive purposes. Researchers from the University of California Santa Barbara borrowed and simplified chemistries that the mussel foot uses to overcome this engineering challenge of wet adhesion to create copolyampholytes, and one-component adhesive systems with potential for employment in nanofabrication protocols.

Mimicking the diving behavior of animals, researchers discovered in 2013 that humans have a similar capacity to lower brain temperature and suppress metabolism for neuroprotection. This has now opened a real possibility of devising means for humans to sustain this state, not unlike the elusive and enigmatic feat of animal hibernation, e.g., lemurs (primates) and bears. This would have profound biomedical implications for healthcare and for treating an unmatched range and diversity of serious life-threatening clinical conditions, and in a fully personalized way, things like stroke, blood-loss, burns, cancer, chronic obesity, epileptic seizures, etc. An experimental trial, recently conducted in Sweden seemingly resulted in a sustainable variant of this state in a human breath-hold diver.

Spider web silk is as strong as the Kevlar used in bulletproof vests. Engineers could in principle use such a material, if it could be reengineered to have a long enough life, for parachute lines, suspension bridge cables, artificial ligaments for medicine, and other purposes. Other research has proposed adhesive glue from mussels, solar cells made like leaves, fabric that emulates shark skin, harvesting water from fog like a beetle, and more. Murray's law, which in conventional form determined the optimum diameter of blood vessels, has been re-derived to provide simple equations for the pipe or tube diameter which gives a minimum mass engineering system. Aircraft wing design and flight techniques are being inspired by birds and bats.

Robots based on the physiology and methods of locomotion of animals include BionicKangaroo which moves like a kangaroo, saving energy from one jump and transferring it to its next jump, and climbing robots, boots and tape mimicking geckos feet and their ability for adhesive reversal. Nanotechnology surfaces that recreate properties of shark skin are intended to enable more efficient movement through water. Tire treads have been inspired by the toe pads of tree frogs. The self-sharpening teeth of many animals have been copied to make better cutting tools. Protein folding is used to control material formation for self-assembled functional nanostructures. The Structural coloration of butterfly wings is adapted to provide improved interferometric modulator displays and everlasting colours. New ceramics copy the properties of seashells. Polar bear fur has inspired the design of thermal collectors and clothing. The arrangement of leaves on a plant has been adapted for better solar power collection. The light refractive properties of the moth's eye has been studied to reduce the reflectivity of solar panels. Self-healing materials, polymers and composite materials capable of mending cracks have been produced based on biological materials.

The Bombardier beetle's powerful repellent spray inspired a Swedish company to develop a "micro mist" spray technology, which is claimed to have a low carbon impact (compared to aerosol sprays). The beetle mixes chemicals and releases its spray via a steerable nozzle at the end of its abdomen, stinging and confusing the victim.

Most viruses have an outer capsule 20 to 300 nm in diameter. Virus capsules are remarkably robust and capable of withstanding temperatures as high as 60 °C; they are stable across the pH range 2-10. Viral capsules can be used to create nano device components such as nanowires, nanotubes, and quantum dots. Tubular virus particles such as the tobacco mosaic virus (TMV) can be used as templates to create nanofibers and nanotubes, since both the inner and outer layers of the virus are charged surfaces which can induce nucleation of crystal growth. This was demonstrated through the production of platinum and gold nanotubes using TMV as a template. Mineralized virus particles have been shown to withstand various pH values by mineralizing the viruses with different materials such as silicon, PbS, and CdS and could therefore serve as a useful carriers of material. A spherical plant virus called cowpea chlorotic mottle virus (CCMV) has interesting expanding properties when exposed to environments of pH higher than 6.5. Above this pH, 60 independent pores with diameters about 2 nm begin to exchange substance with the environment. The structural transition of the viral capsid can be utilized in Biomorphic mineralization for selective uptake and deposition of minerals by controlling the solution pH. Possible applications include using the viral cage to produce uniformly shaped and sized quantum dot semiconductor nanoparticles through a series of pH washes. This is an alternative to the apoferritin cage technique currently used to synthesize uniform CdSe nanoparticles. Such materials could also be used for targeted drug delivery since particles release contents upon exposure to specific pH levels.

References

- Flint, Maria Louise & Dreistadt, Steve H. (1998). Clark, Jack K., ed. Natural Enemies Handbook: The Illustrated Guide to Biological Pest Control. University of California Press. ISBN 978-0-520-21801-7.

- Pauline Pears (2005), HDRA encyclopedia of organic gardening, Dorling Kindersley, ISBN 978-1405308915

- Acorn, John (2007). Ladybugs of Alberta: Finding the Spots and Connecting the Dots. University of Alberta. p. 15. ISBN 978-0-88864-381-0.

- Kaya, Harry K. et al. (1993). "An Overview of Insect-Parasitic and Entomopathogenic Nematodes". In Bedding, R.A. Nematodes and the Biological Control of Insect Pests. CSIRO Publishing. ISBN 978-0-643-10591-1.

- Xuenong Xu (2004). Combined Releases of Predators for Biological Control of Spider Mites Tetranychus urticae Koch and Western Flower Thrips Frankliniella occidentalis (Pergande). Cuvillier Verlag. p. 37. ISBN 978-3-86537-197-3.

- Fry, William E. (2012). Principles of Plant Disease Management. Academic Press. p. 187. ISBN 978-0-08-091830-3.

- Dalley, S. (2002). Mari and Karana: Two Old Babylonian Cities (2 ed.). Gorgias Press LLC. p. 203. ISBN 978-1-931956-02-4.

- Crane, Eva The World History of Beekeeping and Honey Hunting, Routledge 1999, ISBN 0-415-92467-7, ISBN 978-0-415-92467-2, 720pp.

- Howard, Fred (1998). Wilbur and Orville: A Biography of the Wright Brothers. Dober Publications. p. 33. ISBN 978-0-486-40297-0.

- Benyus, Janine (1997). Biomimicry: Innovation Inspired by Nature. New York, USA: William Morrow & Company. ISBN 978-0-688-16099-9.

- "Classical Biological Control: Importation of New Natural Enemies". University of Wisconsin. Retrieved 7 June 2016.

- "The Chinese Scientific Genius. Discoveries and inventions of an ancient civilization: Biological Pest Control" (PDF). The Courier. UNESCO: 24. October 1988. Retrieved 5 June 2016.

- "How to Manage Pests. Cottony Cushion Scale". University of California Integrated Pest Management. Retrieved 5 June 2016.

- "Biological control. Phytoseiulus persimilis (Acarina: Phytoseiidae)". Cornell University. Retrieved 7 June 2016.

- Shapiro-Ilan, David I; Gaugler, Randy. "Biological Control. Nematodes (Rhabditida: Steinernematidae & Heterorhabditidae)". Cornell University. Retrieved 7 June 2016.

- "Conservation of Natural Enemies: Keeping Your "Livestock" Happy and Productive". University of Wisconsin. Retrieved 7 June 2016.

- Gurr, Geoff M. (22 February 2016). "Multi-country evidence that crop diversification promotes ecological intensification of agriculture". Nature Plants. doi:10.1038/nplants.2016.14.

- Wilson, L. Ted; Pickett, Charles H.; Flaherty, Donald L.; Bates, Teresa A. "French prune trees: refuge for grape leafhopper parasite" (PDF). University of California Davis. Retrieved 7 June 2016.

- Capinera, John L. (October 2005). "Featured creatures:". University of Florida website - Department of Entomology and Nematology. University of Florida. Retrieved 7 June 2016.

- "The cane toad (Bufo marinus)". Australian Government: Department of the Environment. 2010. Retrieved 2 July 2016.

- "Moving on from the mongoose: the success of biological control in Hawai'i". Kia'i Moku. MISC. 18 April 2012. Retrieved 2 July 2016.

- "Oldest known archaeological example of beekeeping discovered in Israel". Thaindian.com. 2008-09-01. Retrieved 2016-03-12.

- "Bee Species Outnumber Mammals And Birds Combined". Biology Online. 2008-06-17. Retrieved 2016-03-12.

- "Economic aspects of beekeeping production in Croatia" (PDF). Veterinarski Arhiv. 79: 397–408. 2009. Retrieved 2016-03-12.

Permissions

We would like to thank the editorial team for lending their expertise to make the book truly unique. They have played a crucial role in the development of this book. Without their invaluable contributions this book wouldn't have been possible. They have made vital efforts to compile up to date information on the varied aspects of this subject to make this book a valuable addition to the collection of many professionals and students.

This book was conceptualized with the vision of imparting up-to-date and integrated information in this field. To ensure the same, a matchless editorial board was set up. Every individual on the board went through rigorous rounds of assessment to prove their worth. After which they invested a large part of their time researching and compiling the most relevant data for our readers.

The editorial board has been involved in producing this book since its inception. They have spent rigorous hours researching and exploring the diverse topics which have resulted in the successful publishing of this book. They have passed on their knowledge of decades through this book. To expedite this challenging task, the publisher supported the team at every step. A small team of assistant editors was also appointed to further simplify the editing procedure and attain best results for the readers.

Apart from the editorial board, the designing team has also invested a significant amount of their time in understanding the subject and creating the most relevant covers. They scrutinized every image to scout for the most suitable representation of the subject and create an appropriate cover for the book.

The publishing team has been an ardent support to the editorial, designing and production team. Their endless efforts to recruit the best for this project, has resulted in the accomplishment of this book. They are a veteran in the field of academics and their pool of knowledge is as vast as their experience in printing. Their expertise and guidance has proved useful at every step. Their uncompromising quality standards have made this book an exceptional effort. Their encouragement from time to time has been an inspiration for everyone.

The publisher and the editorial board hope that this book will prove to be a valuable piece of knowledge for students, practitioners and scholars across the globe.

Index

A

Abdomen, 21, 34, 39-43, 45-46, 59, 63-64, 66, 71-73, 75, 79, 82, 94, 103, 108, 110-111, 118-119, 131-133, 136, 143, 145, 163, 180-181, 201-202, 204, 208, 223, 283, 302

Adhesive Traps, 242

Aggregating Pheromones, 158, 252

Air Exposure, 25

Arthropod, 1, 6-7, 14-15, 17, 19, 26-27, 34, 42, 100, 192, 220, 243, 254, 256

Augmentation, 258, 260, 262-263

B

Bee, 22, 34, 53-54, 59, 61, 77, 83, 110, 150, 159-163, 165-176, 206, 217, 228, 238, 270, 273-280, 282-289, 292, 294, 304

Beetle, 4, 8-9, 23, 25, 32, 34-35, 51, 57-58, 63-75, 77-80, 82-89, 99, 119, 158, 220, 228, 243, 260-263, 265, 267-269, 302

Beneficial Insects, 32, 158, 260, 263

Biological Pest Control, 60, 101, 158, 177, 213, 217, 258, 260, 263, 268, 270, 303-304

Biomimetics, 199, 258, 297-298, 300

Bottle Trap For Insects, 244

Butterflies, 2-4, 23, 50, 52, 59, 61, 128-131, 134-136, 138-141, 143-148, 150-153, 155-159, 171, 214, 233-235, 238-239, 246, 250, 265, 300

C

Caddisfly, 214-217

Circulatory System, 46-47, 74, 136, 204

Coevolution, 38, 141, 155, 161

Coleoptera, 2, 8, 22, 25, 39-40, 52, 59, 63-67, 69-71, 76, 82, 84-85, 128, 236

Coleopterology, 4, 8-9

Communal Bees, 165

Conservation, 10, 165, 258, 260, 263, 304

Contemporary Myrmecologists, 12

D

Digestive System, 43, 74, 135

Diversity, 10, 38-39, 61, 64-67, 70, 81, 103, 105-107, 114, 129-130, 133, 155, 160, 179, 191, 203, 217, 219, 221, 233, 238, 302

E

Ecological Roles, 35, 97, 195, 218, 233

Economic Entomology, 8, 29-30

Eusociality, 10, 85, 164, 217

Evolution Of Hive Designs, 277

Exoskeleton, 34, 36, 42, 49, 55, 64, 66, 71, 180, 184, 204, 222

Extremities, 72, 180

F

Fixed Comb Hives, 280-281

Flies, 4, 15, 19-22, 24-26, 35, 37, 44, 48, 60, 62, 82, 87, 114, 120, 141, 146, 150, 157, 160, 171, 173, 194, 206, 214, 216, 242, 246, 251, 265, 267, 269

Flight, 34, 37, 42, 47, 55-56, 73, 76, 79, 96, 109-111, 114-116, 125, 133, 139, 143-144, 146, 148-149, 163, 168-169, 172, 181, 184, 204, 208, 234-235, 241, 243, 247, 250, 257, 286, 298, 302

Flight Interception Trap, 247, 250

Flying Insect Traps, 243

Foraging, 51, 54, 109-110, 115-116, 120, 183, 185, 188, 190, 195, 199, 222-223, 243, 287

Forensic Entomology, 8, 17-22, 25-26, 29

Fossil Record, 66-68, 152, 155, 218-219

H

Haplodiploid Breeding System, 163

Harmful Insects, 31

Hemiptera, 34, 52, 64, 90, 93-94, 96, 98, 100-101, 103, 232, 267

Herbivores, 38, 97, 140, 179, 189, 259

I

Importation, 258, 260, 304

Insect Ecology, 57, 231, 233, 235, 237, 239

Insect Hotel, 237-239

Insect Trap, 158, 241-244, 252

Interdisciplinary Application, 10

Invertebrate, 12, 20, 120, 146, 175, 265

J

Jumping, 73, 95-96, 101, 142, 190-191, 201, 204, 224

K

Killing Jar, 248

L

Larvae, 20-24, 26, 28-29, 35, 43, 47, 52, 62, 67, 70, 74, 76-78, 80-81, 83-84, 86-88, 99, 128-129, 132-133, 135, 138-142, 146-147, 149-151, 153-160, 168, 172, 176, 181, 183, 188, 191-193, 196-197, 214-217, 220-221, 223-224, 226-228, 238, 261, 264-268, 270, 275, 277, 287-288, 290-292, 296

Lepidoptera, 2, 23, 39-40, 50, 52, 59, 128-133, 136-139, 143-144, 146-148, 151-158, 171, 214, 216, 226, 235

Lepidopterans In Diapause, 138

Lifecycle, 77, 131, 135, 138-140, 150, 205, 216

Light Traps, 242, 256

Locomotion, 55-57, 79, 95, 108, 115, 143, 190, 199, 302

M

Major Insect-born Disease, 16

Malaise Trap, 243, 247, 249-250

Marangoni Propulsion, 57, 96

Mating, 18, 46-47, 64, 71, 76, 81, 85, 90, 111, 128, 133, 136, 138-140, 143, 145, 154, 164, 167-168, 183-184, 190, 225, 236, 252, 254, 274-275, 286-287, 292

Medical Entomology, 8, 14, 33

Metamorphosis, 34, 39, 49-50, 64, 76, 94, 104, 110-111, 128, 138, 154, 167-168, 182, 200, 205, 286, 292

Migration, 104, 137, 145, 233-236, 240

Moisture Levels, 24

Morphology, 1, 27, 38, 40, 64, 66, 69-70, 73, 130, 134-135, 152, 160, 180, 182

Moth Trap, 250-251

Mounds, 106-107, 114, 119-122, 125-128, 301

Mouthparts, 41, 44, 71-72, 78, 84, 90, 93, 97-98, 131-132, 138, 140, 142, 155, 162-163, 201, 203, 223

Myrmecology, 4, 8-11, 33

N

Nervous System, 43, 74, 180, 203-204

O

Odonata, 31, 63, 236, 240

Orientation, 54, 107, 122, 127, 144, 235, 286

Orthoptera, 41, 44, 200, 202, 236

P

Parasites, 60, 70, 99-100, 102, 114-115, 121, 141, 150, 167, 171-173, 191, 194, 201, 206, 217, 227, 292

Parasitic Insects, 51, 239

Parasitoids, 58, 81, 101, 150-151, 201, 206, 220-221, 226, 259-260, 263, 265-266, 268

Parental Care, 54-55, 77, 80, 101, 227

Pathogens, 15, 114-115, 117-118, 121, 172, 187, 206, 259-260, 266-268

Personal Pests, 15

Phylogeny, 35, 37, 69, 152-153, 162, 202, 218

Physiology, 1-2, 40, 232, 302

Pitfall Trap, 253

Pollination, 38, 57, 59-61, 83, 148, 159-161, 166, 175, 193, 220, 225, 233, 238

Polymorphism, 48, 59, 136-137, 148, 181-182, 234

Predation, 51, 53, 58-59, 81-82, 97-98, 146, 148, 171, 192, 216, 227, 233, 258-259, 269-270

Predators, 22, 32, 35, 38, 52, 58-59, 61-62, 75, 80-83, 85, 87, 90, 93, 98, 100-102, 110, 113-115, 118, 120-121, 123, 141-143, 146-148, 151, 160, 167, 172-173, 175, 178-179, 187, 189, 192, 201, 205-206, 208, 214, 218, 220, 227-229, 232, 239, 253, 259-260, 263-265, 269-270, 282, 303

Pupa, 27-28, 50, 76, 78, 128, 131, 138, 141-142, 154-155, 168, 183, 215, 226, 288

R

Reproductive System, 45-46, 134

Respiratory System, 46, 56, 74, 136

S

Scanning Electron Microscopy, 26-27

Scorpionflies, 20

Segmentation, 41, 72

Sericulture, 33, 258, 294-295

Skating, 95

Solitary Bees, 166-168, 238-239

Sound Production, 52, 94, 103, 202

Specialized Organs, 50, 75

Sternorrhyncha, 93-94, 96-99, 101

Submerged Corpses, 22, 24

Sun Exposure, 24

Swimming, 34, 57, 73, 95

T

Taxonomy, 38-39, 65, 105, 155, 178, 202, 218

Termite, 34, 104-115, 117-128, 301

The Cockroach, 15, 48, 104-106

The Housefly, 15

Top-bar Hives, 281

Trapping Pit, 253

Tullgren Funnel, 254-255

V

Veterinary Entomology, 14

Viruses, 114-115, 207, 266-268, 292, 302-303

W

Wasp, 22, 34, 43, 50-51, 80, 83, 139, 145, 148-151, 157, 160, 163, 177, 190, 194, 217-231, 243, 259, 265-266

Wasps, 2, 4, 13, 24, 37, 48-49, 51, 53, 55, 59-61, 82, 99, 104, 110, 114, 142, 146, 150, 159-160, 162-164, 171, 173, 177-179, 194, 206, 217-230, 238-239, 242, 244-246, 251, 262, 265, 268-269

Wing Structure, 93

Wings, 21, 35, 37, 39, 41-42, 49, 55-56, 69, 72-73, 75, 79, 94, 103-106, 108-109, 115, 125, 128-129, 131-133, 136-138, 141-143, 146-147, 153, 155, 162-163, 169, 181, 184, 190, 195, 201, 204-205, 208, 214, 223, 230, 234, 239, 245-246, 287, 300-302

www.ingramcontent.com/pod-product-compliance
Lightning Source LLC
Chambersburg PA
CBHW061930190326
41458CB00009B/2711

* 9 7 8 1 6 3 5 4 9 1 0 8 1 *